3.95

Think Metric Now!

A Step-by-Step Guide to Understanding and Applying the Metric System

Paul J. Hartsuch, Ph.D.
Member,
Metric Association, Inc.

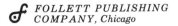 *FOLLETT PUBLISHING COMPANY, Chicago*

Manufactured in the United States of America.

Library of Congress Catalog Card Number: 73-91583

ISBN: 0-695-80449-9

456789/82818079787776

PREFACE

I, of course, hope the reader will read this book carefully and learn how to use the metric system, and maybe even have a little fun while doing it. But, to be honest, my main object was to write a book that would be purchased by thousands of people, so I will have lots of money with which to purchase kilograms of potatoes, many thousand grams of butter, liters and liters of milk, and be able to drive my car thousands of kilometers. Your cooperation in the realization of this aim will be greatly appreciated.

CONTENTS

PART ONE

1. Weight   *3*
   Metric Weights   3
   Converting One Metric Weight to Another   7
   General Rule for Moving Decimal Point   8
   Metric Prefixes   9
   Key Weight Conversions   10
   Problems   11

2. Length   *13*
   Metric Lengths   13
   Adding Metric Lengths   17
   Kilometers   19

Dividing Metric Lengths 20
Reviewing Units of Length 21
Problems 22

3. How About Areas *24*
   Finding Area in Square Centimeters 25
   Finding Area in Square Meters 26
   Finding Area in Ares and Hectares 27
   Finding Area in Square Kilometers 28
   Consistency of Units 28
   Relation of Metric Areas 29
   Key Area Conversions 31
   Problems 31

4. Volume *34*
   Liters 35
   Milliliters and Cubic Centimeters 37
   Cubic Meters 39
   Irregular Volumes 40
   Reviewing Volumes 40
   Problems 41

5. Connecting Volume and Weight *43*
   Density 44
   Finding Weight from Volume 45
   Finding Volume from Weight 47
   Problems 48

6. How Hot or Cold Is It? *50*
   Reading a Celsius Thermometer 51

Converting Celsius Reading to Fahrenheit 52
Problems 54

7. Learning to "Think Metric" *55*

PART TWO

8. Metric Prefixes *61*

9. Metric-Metric Conversion *64*

10. Converting Celsius and Fahrenheit Temperature
    Readings *69*

11. Calculating Areas and Volumes *72*
    Problems 76

12. English-Metric Conversion *77*

13. An Easy Way To Solve Conversion Problems *82*
    Problems 90

14. Using Powers of Ten *92*
    Problems 98

    Answers to Problems 99

    Index 114

PART ONE

1

Weight

You must want to learn more about the metric system or you wouldn't read even the first page of this book. So let's forget that the United States is the last big country to make the switch to metric, the long story about how the metric system came into existence, and all the other stuff that fills up space in newspapers and magazines. Instead, let's begin at once to get acquainted with the comparatively few common metric units, and to try to "Think Metric."

METRIC WEIGHTS

In the non-metric system we have been using in the United States, the pound is one unit of weight. For smaller weights there are ounces, and there are 16 ounces in one pound. For really large weights, there is the ton. For most people, a ton consists of 2,000 pounds, although to be correct this is a "short ton." A "long ton" is 2,240 pounds. There are other units such as troy ounces, drams, scruples, and grains, but relatively few people use them.

It is convenient in any system of measurement to have small, medium, and large units of weight and this is true of the metric system as well. One metric unit of weight is called a *gram,* and is abbreviated as "g." A gram is a much smaller unit than an ounce: it takes a little over 28 grams to make 1 ounce.

To give you a feeling for how much a gram weighs, heft some of the coins in your pocketbook. Here are the approximate weights, in grams, of common coins:

> A dime—a little over 2 grams
> (actually 2.2 g.)
> A penny—about 3 grams
> A nickel—about 5 grams
> A quarter—5.65 grams
> A half dollar—11.4 grams

A bigger unit of weight is the *kilogram* (kg). A kilogram consists of 1,000 grams. Now you can begin to see why the metric system is so wonderful. Everything in it goes by 10s, 100s, or 1,000s, which makes arithmetical calculations much easier. A kilogram is bigger than a pound. It takes 2.2 pounds to equal 1 kilogram.

The biggest metric unit of weight is the metric ton (t). A metric ton consists of 1,000 kilograms. A metric ton is almost the same weight as an English short ton. To make a rough conversion from metric tons to

Quick Reference Chart for Metric-Metric Conversions

WEIGHT

1 gram (g) = 1,000 milligrams (mg)
1 kilogram (kg) = 1,000 grams (g)
1 metric ton (t) = 1,000 kilograms (kg)

LENGTH

1 centimeter (cm) = 10 millimeters (mm)
1 meter (m) = 100 centimeters (cm)
1 meter (m) = 1,000 millimeters (mm)
1 kilometer (km) = 1,000 meters (m)

VOLUME

1 liter (l) = 1,000 milliliters (ml)
1 liter (l) = 1,000 cubic centimeters (cm³)
1 milliliter (ml) = 1 cubic centimeter (cm³)
1 cubic meter (m³) = 1,000 liters (l)

AREA

Width in centimeters (cm) × length in centimeters (cm) = area in square
   centimeters (cm²)
Width in meters (m) × length in meters (m) = area in square meters (m²)
Width in kilometers (km) × length in kilometers (km) = area in square
   kilometers (km²)
1 hectare (ha) = an area 100 meters (m) wide and 100 meters (m) long
1 hectare (ha) = 10,000 square meters (m²)
100 hectares (ha) = 1 square kilometer (km²)
1 are (a) = an area 10 meters (m) wide and 10 meters (m) long
         = 100 square meters (m²)
1 hectare (ha) = 100 ares (a)

RELATION BETWEEN VOLUME AND WEIGHT

1 liter (l) of water (at 4°C) = 1 kilogram (kg)
1 milliliter (ml) or 1 cubic centimeter (cm³) of water (at 4°C) = 1 gram (g)

MEANING OF PREFIXES

Prefix	Means	
centi-	$\frac{1}{100}$	1 centimeter = $\frac{1}{100}$ of a meter or there are 100 cm in 1 meter
milli-	$\frac{1}{1000}$	1 milligram (mg) = $\frac{1}{1000}$ of a gram or there are 1,000 mg in 1 gram
kilo-	1,000	1 kilometer = 1,000 meters

[NOTE: Don't bother to look at these now. They are for quick reference after you have read the book. See page 100 for summary of key conversions between metric and English units.]

English short tons, take the number of metric tons and add 10%. For example, 100 metric tons are equivalent to about 110 English short tons. Similarly 200 metric tons are equal to about 220 English short tons.

For everyday use, this is all you need to know about metric units of weight. Here are the measurements again:

A gram = about $\frac{1}{28}$ of an ounce
A kilogram = 1,000 grams
  = 2.2 pounds
A metric ton = 1,000 kilograms
  = 1.1 short tons

The size of these units in relation to English units has been indicated, but don't try to memorize these conversions. The idea is to work entirely in the metric system instead of continuously making conversions to the old units.

When scales are converted, you won't have to worry. You put whatever you are measuring—paper, fruits, vegetables—on the scale, and read the weight. If it is a scale for light objects, it will give you the weight in grams. If it is a scale for heavier objects, it will read in kilograms—and likely, in tenths of kilograms, such as 5.7 kilograms. A gram-reading scale may also have ten subdivisions for each gram. For example, you can read a value such as 34.8 grams. You won't have to worry about fractions in the metric system and you can forget such figures as $\frac{1}{4}$, $\frac{3}{8}$, and $\frac{7}{32}$.

Converting one metric weight to another is easy—once you know how. It merely involves moving the decimal point the right number of places either left or right. There are 1,000 grams in a kilogram. To convert grams to kilograms, move the decimal place three places to the left. For example, 350 grams are 0.350 kilogram. Likewise, 4,760 grams are 4.760 kilograms. Whole numbers usually do not show the decimal point, but they could be written 350. and 4,760. grams.

If a conversion figure of 1,000 is involved, the decimal point is moved three places. It is easy to know which way to move it. You know that a kilogram is much bigger than a gram, so you must move the decimal point in the gram figure three places to the left to give the corresponding weight in kilograms.

The reverse is true to convert kilograms to grams. In a given quantity there are a lot more grams than kilograms. Therefore, you move the decimal point three places to the right. For instance, 2.1 kilograms are 2,100 grams. Note that you have to add two more zeros. If it is easier for you, use 2.100 kilograms. This is exactly the same amount as 2.1 kilograms.

The same procedure is used to convert kilograms to metric tons, or the reverse. This is because there are 1,000 kilograms in one metric ton. Thus 750 kilograms equals 0.750 metric ton, and 45,200 kilograms equals 45.200 metric tons. Going the other way, 10.5 metric tons are equal to 10,500 kilograms. Note that you have to add two more zeros.

At this point, you should practice making these conversions. Write any figure in grams, and then convert it to kilograms. Then write a fairly small figure in kilograms, and convert it to grams. Do the same thing for kilograms to metric tons and metric tons to kilograms.

The metric units of weight we have discussed so far differ by a factor of 1,000. Other units, to be discussed later, differ by a factor of 10 or 100. For example, there are 10 millimeters to a centimeter in measuring length. And there are 100 centimeters in a meter. To convert units that vary by a factor of 10, the decimal point is moved only one place. If units vary by a factor of 100, the decimal point is moved two places. Here is a summary of how much to move the decimal point:

Units differ by factor of	Number of places decimal point is moved
10	1
100	2
1,000	3
10,000	4

There will be many more examples of moving the decimal point as you proceed, and soon you will be moving that little decimal point with the greatest of ease. It is the ability to do this that makes the metric system the easiest one that was ever devised.

8 *Think Metric Now!*

The relation between metric units of weight has been given. There are 1,000 grams in a kilogram, and 1,000 kilograms in a metric ton. Three metric units of length were just mentioned—the meter, the centimeter and the millimeter. While you *must* learn these conversion factors, remember that the prefix of some units tells its relation to other units. This makes it easy to remember them. The prefix "centi" means $\frac{1}{100}$, and "milli" means $\frac{1}{1000}$. Thus a centimeter is $\frac{1}{100}$ of a meter. To put it another way, there are 100 centimeters in a meter. A millimeter is $\frac{1}{1000}$ of a meter. Or there are 1,000 millimeters in a meter.

Going in the other direction, the prefix "kilo" means 1,000. Thus a kilogram is 1,000 grams. And a kilometer is 1,000 meters.

There are some other prefixes in the names of metric units, but you will probably never use them. Here is a summary of the meanings of the commonly used prefixes:

Prefix	Meaning
centi-	$\frac{1}{100}$
milli-	$\frac{1}{1000}$
kilo-	1,000

(For those interested, all the metric prefixes are given in Part II, Chapter 8.)

It is desirable to leave the English system of weights behind just as fast as possible and to begin to think in metric weights. But at the start this is naturally quite difficult to do. A few conversions between the two systems are needed as a crutch. You don't need a complicated conversion table; in fact, a few "key" conversions will suffice until you get a feeling for the new system. I suggest that you keep only the following in mind:

> 30 grams are a little more than an ounce.
> 500 grams are a little more than a pound. (450 grams come much closer.)
> 1 kilogram is 2.2 pounds.
> 1 metric ton is about 10% larger than an English short ton.

If you learn these key conversions, you can easily go up or down from them to figure approximately how much something weighs. Thus if you buy 500 grams of butter, you are getting a little over a pound, and 250 grams is a little over a half pound. If you buy 5 kilograms of potatoes, you are getting 5×2.2, or 11 pounds. If you purchase 50 grams of a small item, it is a little over $1\frac{1}{2}$ ounces. I have used this rough kind of conversion on trips to Europe and Japan, where the metric system is used, and it works fine. Who needs to know conversions to the second decimal place?

Incidentally, there is one more metric unit of weight. It is called a milligram. From what you have just learned, you may deduct that a milligram must be $\frac{1}{1000}$ of a gram, or that there are 1,000 milligrams in a gram. Since a gram is $\frac{1}{28}$ of an English ounce, and a milligram is $\frac{1}{1000}$ of a gram, it is evident that a milligram is a very small unit of weight. It is used by people who must weigh very small quantities of materials. The average person never uses it.

PROBLEMS

It is one thing to say that you understand the material that has been presented about metric units of weight, but the real test is to see if you can use it properly. The problems that follow will test you on this.

Before you attempt these problems, you should memorize all the relationships of the weights. You should know, for example, that 1 kilogram is equal to 1,000 grams. Work all of the problems before referring to the answers listed in the back of the book, page 99. *Please* don't peek until after you have worked them.

1. Convert:
   a. 550 grams to kilograms
   b. 55 grams to milligrams
   c. 247 milligrams to grams
   d. 5.48 kilograms to grams
   e. 548 kilograms to metric tons
   f. 25.8 metric tons to kilograms

2. Add:
   a. 2.42 kilograms + 370 grams. Give answer in kilograms. Also give answer in grams.
   b. 350 kilograms + 5.15 metric tons. Give answer in kilograms. Also give answer in metric tons.
   c. 2.84 grams + 250 milligrams. Give answer in grams. Also give answer in milligrams.

3. Is 500 grams of bread more than, equal to, or less than one pound of bread? Why?

4. If a man weighs 176 pounds, what is his approximate weight in kilograms?

5. One figure for a daily basic diet is 25 calories per kilogram of body weight. How many calories would this be for a person weighing 160 pounds?

6. One package of raisins contains 170 grams and sells for 15 cents. Another contains 425 grams and sells for 47 cents. Which is the best buy? Prove it by some method.

7. If potatoes sell for 15 cents per pound, what should the selling price be for one kilogram of potatoes?

2

Length

You have, I hope, learned something about metric units of weight. If you haven't, go back over the material in Chapter 1 until you have mastered it. Don't try to read this book as if it were a novel. Take it one section at a time and when you think you have learned that section, go on to the next one.

One unit of length in the metric system is the *meter* (m). (Sometimes it is spelled "metre," but we will use "meter.") A meter is 39.37 inches, or just a little over 3 inches longer than a yard. A meter is roughly 10% longer than a yard. If a room is 6 meters long then it is approximately 6.6 yards long (10% of 6 is 0.6, and 6 plus 0.6 is 6.6).

In Chapter 1 you learned the common metric prefixes. So you should be able to state that a *centimeter* (cm) is $\frac{1}{100}$ of a meter, or that there are 100 centimeters in one meter. Also that a *millimeter* (mm) must be $\frac{1}{1000}$ of a meter, or that there are 1,000 millimeters in one meter. Finally, that a *kilometer* (km) must be 1,000 meters. These are the common metric units of length.

Short lengths are usually measured in centimeters. A centimeter is not as long as an inch. An inch is about $2\frac{1}{2}$ centimeters, so a foot is about 30 centimeters. A section of a 20-centimeter ruler is shown in Figure 1. You will notice that each centimeter is divided into ten parts. Each of these is 0.1 cm. Actually, each of these small divisions is a millimeter, but for measurements it is better to think of them as tenths of a centimeter.

It is desirable to read a length rapidly to the nearest tenth of a centimeter. One thing that helps is that the tenth mark halfway between centimeter marks is a little longer than the other tenth marks. So you know at once that this is 0.5 (five tenths). If a measurement comes one mark less than 0.5, it is read as 0.4. If one mark less than the next centimeter mark, it is read as 0.9. This is shown in Figure 2. *Don't* count these tenth marks from one up. With a little practice, any tenth mark can be read immediately, without counting. You can even read 0.7 without counting. It is near the middle between 0.5 in the center and the next centimeter mark, but a little closer to 0.5. If the mark is in this

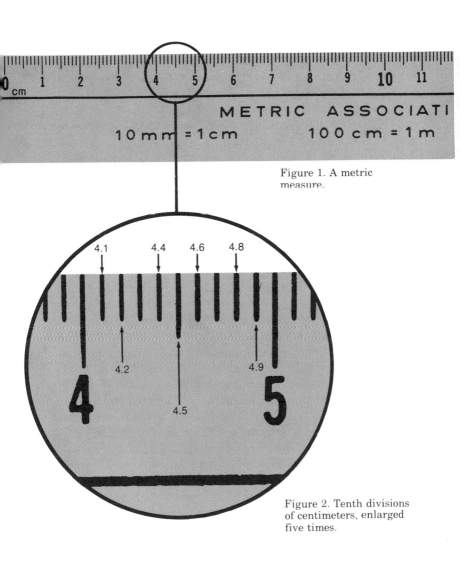

Figure 1. A metric measure.

Figure 2. Tenth divisions of centimeters, enlarged five times.

Length 15

same region, but a little closer to the next centimeter mark, it is read as 0.8. This same thing holds true for readings of 0.2 and 0.3, on the left side of the 0.5 mark.

For practice, a number of objects should be measured with a 20- or 30-centimeter ruler, to give readings such as 12.8 cm., or 15.3 cm. This is much easier than reading $\frac{3}{8}''$, $\frac{9}{16}''$, etc.

Somewhat longer lengths are measured in meters. At this point, you really ought to obtain a meter stick, so you can make some measurements with it. You will notice that a meter stick is divided into 100 equal parts, and, of course, each of these parts is a centimeter.

This means you can measure a distance that is not an even number of meters. Taking the length of a room, suppose that you lay down the meter stick six times before finding yourself near the opposite wall. You can measure the remaining distance in centimeters from your last mark on the floor to the wall. The length of the floor might be 6 meters, 35 centimeters.

I have mentioned a meter stick, since it is a convenient length. It is also possible to get metric lengths in several ways, such as a folding rule, a flexible tape, or a shorter, 20-centimeter ruler. The type of tool does not make any difference, of course, as a centimeter is the same length on all of them.

To learn to "Think Metric" in units of length, you should use whatever metric measuring device you have, and measure a lot of things—the width and length of a table, the width and height of a door, or the dimensions of various sheets of paper.

The metric system has been used for a long time in photography. You may be familiar with 35-millimeter color transparencies. Actually they are 35 millimeters long and only 23 millimeters wide—or 3.5 by 2.3 centimeters. Looking at a 35-millimeter slide will help you develop a feeling for millimeter lengths. The cardboard in which the slide is mounted is 50 millimeters by 50 millimeters. It also is 5 centimeters by 5 centimeters.

The focal length of camera lenses is also given in metric units. The typical lens of a 35-millimeter camera has a focal length of 50 millimeters (5.0 centimeters). A wide angle lens may have a focal length of 27 to 28 millimeters, while the focal length of a telephoto lens is much longer and may go up to 1,000 millimeters (100 centimeters or 1 meter).

ADDING METRIC LENGTHS

One rule of any system of measurement is that if you want to add two or more figures, they must be expressed in the same units. If you want to add 3 feet and 17 inches, you either have to convert the feet to inches and get the sum of the two numbers in inches or convert inches to feet and get the sum in feet.

The metric system is no different. But the conversion from one unit to another is much simpler. For example, 3 meters are 300 centimeters. And 45 centimeters are 0.45 meter. (Remember that for a factor of 100, the decimal point must be moved two places one way or the other.)

The problem that follows shows how to add two lengths. One length is 2 meters, 67 centimeters, and the other is 3 meters, 24 centimeters. First let's change both amounts to centimeters:

$$2 \text{ m } 67 \text{ cm} = 267 \text{ cm}$$
$$3 \text{ m } 24 \text{ cm} = \underline{324 \text{ cm}}$$
$$591 \text{ cm}$$

Next change both amounts to meters:

67 cm = 0.67 m so the first length is   2.67 m
24 cm = 0.24 m so the second length is   3.24 m
The total is   5.91 m

Both answers come out to 591, so maybe we have done both of them correctly. To check this, convert 591 centimeters to meters. The answer is 5.91 meters. And if you convert 5.91 meters to centimeters, we get 591 centimeters. Good! We did both methods correctly.

There is a way of adding lengths such as those given above and not changing them either to centimeters or to meters. For example:

$$\begin{array}{r} 3 \text{ m} \quad 24 \text{ cm} \\ + 5 \text{ m} \quad 35 \text{ cm} \\ \hline 8 \text{ m} \quad 59 \text{ cm} \end{array}$$

$$
\begin{array}{r}
3 \text{ m} \quad 70 \text{ cm} \\
+ \ 5 \text{ m} \quad 82 \text{ cm} \\
\hline
8 \text{ m } 152 \text{ cm} = 9 \text{ m } 52 \text{ cm}
\end{array}
$$

In the second example, 152 cm = 1 m 52 cm. So the proper way to record the answer is 9 m 52 cm.

KILOMETERS

Another unit of length is the *kilometer* (km), which is equal to 1,000 meters. A kilometer is roughly 0.6 mile. For example, if a speed limit sign on a highway says 100 kilometers per hour, this is a speed limit of about 60 miles per hour ($100 \times 0.6 = 60$).

Several years ago, I picked up a Volkswagen in Germany and drove it around Europe. Since it was designed to be used in the United States, it was equipped with a speedometer that read "miles per hour." But all the road signs in Europe are "kilometers per hour." I had to multiply the kilometer figure by 0.6 to find out how fast I could go in miles per hour without exceeding the speed limit. For example, if a sign read 60, then I had to slow down to 60×0.6, or 36 mph.

There is a factor of 1,000 between kilometers and meters, so it is necessary to move the decimal point three places to convert one unit to another. Thus 5.6 kilometers is 5,600 meters. And 850 meters is 0.850 kilometers. An 800-meter race is 0.8 kilometer, and 0.8 kilometer $\times$ 0.6 is 0.48 mile. So an 800-meter race is slightly less than a half-mile race.

With the metric system it is easy to find the exact center of a distance—be it the center of a wall or the center of one side of a sheet of paper. Suppose a length is 43.8 centimeters. The center of this length is $\frac{1}{2}$ of 43.8, or 21.9 centimeters. It is almost as easy if you want $\frac{1}{3}$ or $\frac{1}{4}$ of this length. You just divide by 3 or by 4.

If you need to find the center of a length such as 2 meters, 35 centimeters, it is necessary either to convert meters to centimeters or centimeters to meters. It doesn't matter which conversion you do. Here are the two methods:

A. 2 meters are 200 centimeters. The length is 200 plus 35, or 235 centimeters. Half of this is 117.5 centimeters. Since there are 100 centimeters in one meter, the distance can be expressed as 1 meter, 17.5 centimeters.

B. 35 centimeters are equal to 0.35 meter. This length can be expressed as 2 plus 0.35 or 2.35 meters. Half of this is 1.175 meters, or 1 meter plus 0.175 meter. But 0.175 meter is 17.5 centimeters. So this half distance again comes out to be 1 meter, 17.5 centimeters.

Calculations of this kind require moving the decimal point before the two numbers are added, but this is a lot easier than finding the center of 4 feet, $3\frac{5}{8}$ inches in the English system.

Briefly, here are metric units of length again:

The standard length is the meter.

A centimeter is $\frac{1}{100}$ of a meter. There are 100 centimeters in one meter.

A millimeter is $\frac{1}{10}$ of a centimeter or $\frac{1}{1000}$ of a meter. There are 10 millimeters in one centimeter, or 1,000 millimeters in one meter.

A kilometer is 1,000 meters.

Try to get away from converting metric lengths into English units as soon as possible. Until you can do this, here are some approximate conversion factors that you can use:

2.5 centimeters are about 1 inch. (An easy way to convert centimeters to inches is to divide by 10 and multiply by 4. Thus 90 cm = 9 × 4 = 36 in.)

30 centimeters are about 1 foot.

1 meter is slightly longer than 1 yard. Add 10% to the length in meters to obtain length in yards.

1 kilometer is about 0.6 mile. Multiply any number of kilometers by 0.6 to get a distance in miles.

This is all you need to know about metric units of length. Metric lengths are also used to calculate areas and volume. This will be covered in Chapters 3 and 4.

PROBLEMS

Try to have metric lengths firmly in mind so you can work these problems without referring back to the material in the chapter.

1. Convert:
   a. 357 millimeters to centimeters
   b. 357 millimeters to meters
   c. 42.8 centimeters to millimeters
   d. 42.8 centimeters to meters
   e. 4.98 meters to centimeters
   f. 730 meters to kilometers
   g. 8.13 kilometers to meters

2. About how many miles are 60 kilometers?

3. About how many miles are 3,400 meters?

4. One of the Olympic races is 400 meters. About how many yards is this?

5. Add:

   2 m 60 cm
   + 3 m 75 cm

   Give answer in meters
   and also in centimeters.

6. Subtract:

   4 m 15 cm
   − 2 m 85 cm

   Give answer in meters
   and also in centimeters.

7. A board measures 2 m 31 cm. If it is to be cut into 3 equal lengths, how many centimeters should each one be? If it is to be cut in the exact center, how many centimeters should be measured from one end?

8. In Los Angeles, Joanna Lynn appeared on several television programs in the role of "Miss Metric." Her "vital statistics" are: (a) 89-58-89 cm; (b) 170 cm tall; and (c) weight, 53 kg. Calculate (a) in inches, (b) in feet and inches, and (c) in pounds. Use approximate conversion figures as given in this chapter.

9. If two cities are 200 kilometers apart, about how many miles are they apart?

10. A car speedometer reads 110 kilometers per hour. What is the approximate speed of the car in miles per hour?

Answers to problems begin on page 102.

3

How About Areas?

Area is the surface included within a set of lines, such as a square, a rectangle, a circle, a triangle, or any given shape. If it is a simple shape, such as a square or a rectangle, the area is obtained by multiplying the length of one side by the length of the other side. In this chapter you will measure simple areas. In Chapter 11 you will learn how to obtain the area of other shapes, such as a circle or a right triangle.

No matter what the shape of an area, it is calculated in square measure. For instance, if a wall is 9 feet high and 13 feet long, its area is 9×13, or 117 square feet.

The method for finding area is exactly the same in the metric system, except the units are metric rather than English. Small areas are usually expressed in square centimeters, larger areas in square meters, and very large areas in square kilometers. Then there are two units, called *ares* and *hectares* that are used to measure land areas.

FINDING AREA IN SQUARE CENTIMETERS

A square centimeter (cm²) is a square that is one centimeter wide and one centimeter long. A square centimeter is smaller than a square inch. (See Figure 3.)

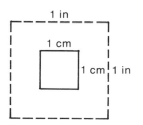

Figure 3. Comparison of a square centimeter and a square inch (actual size).

In fact, about $6\frac{1}{2}$ square centimeters equal one square inch. To find the number of square centimeters in a small area, measure the width of one side, in centimeters, and the length of another side, in centimeters, and multiply the two. If a rectangle is 20 centimeters wide and 30 centimeters long, its area is 20 × 30, or 600 square centimeters.

Of course, the numbers to be multiplied do not have to be whole numbers. One sheet of paper might measure 21.6 centimeters in width and 27.9 centimeters in length. Its area is 21.6 × 27.9, or 602.64 square centimeters.

Note: If 4 is multiplied by 4, the answer can be expressed as 4². Similarly, if a length in cm is multiplied by a width in cm, the answer is cm². So this is the abbreviation for "square centimeters."

FINDING AREA IN SQUARE METERS

In Chapter 2 you learned that a meter is 39.37 inches, compared to 36 inches in a yard. So a square that is one meter on each side must be a little bigger than a square that is one yard on each side. That is, a square meter (m^2) is a little bigger than a square yard (yd^2). This comparison is shown in Figure 4.

Figure 4. Comparison of a square yard and a square meter (not drawn to measure).

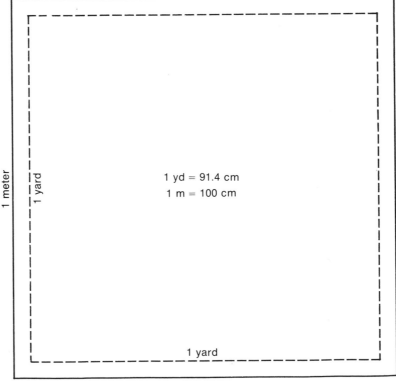

1 meter

1 yard

1 yd = 91.4 cm
1 m = 100 cm

1 yard

1 meter

To find an area in square meters, the same method is used as described above for square centimeters— except each side is measured in meters. If an area is 4 meters wide and 7 meters long, it contains 4×7, or 28 square meters.

An *are* is an area 10 meters wide by 10 meters long. So the area of 1 are is 10×10 or 100 square meters.

A hectare is an area 100 meters wide by 100 meters long. So the area of 1 hectare is 100×100 or 10,000 square meters. Hectares and ares are used mostly for measuring land. One hectare is equal to about $2\frac{1}{2}$ acres. There are 100 ares in 1 hectare.

To find an area in hectares, it is necessary to measure the width and length in meters, multiply the two to get square meters, and then convert this into hectares. This is easy to do in the metric system. All that is necessary is to move the decimal the right number of places. You will recall that when two metric units differ by a factor of 1,000, the decimal point is moved three places. There is a factor of 10,000 between hectares and square meters, so the decimal point must be moved four places to convert one to the other.

To illustrate this, consider a piece of land that is 500 meters wide and 750 meters long. You want to find how many hectares it contains. First multiply 500 meters by 750 meters to give 375,000 square meters. Now move the decimal four places to the left to obtain 37.5 hectares.

Most readers will not be concerned with land areas. It will probably be a long time before land in this country will be recorded in hectares instead of acres.

FINDING AREA IN SQUARE KILOMETERS

Very large area can be expressed in square kilometers. A square kilometer is an area 1,000 meters wide by 1,000 meters long. For example, if a big land area is 3 kilometers wide and 4 kilometers long, it contains 3 × 4, or 12 square kilometers. (There are 100 hectares in a square kilometer.)

Since a kilometer is about 0.6 mile, a square kilometer represents a smaller area than a square mile. It takes about 2.7 square kilometers to equal 1 square mile.

CONSISTENCY OF UNITS

In Chapter 2 it was emphasized that amounts to be added or subtracted must be in the same units. You add 10 centimeters and 32 centimeters to get 42 centimeters. Or you subtract 3.25 meters from 4.60 meters to get 1.35 meters. In the first example, both figures were in centimeters; in the second, both were in meters.

The same thing is true for areas. For example, if width is measured in centimeters, then length must also be measured in centimeters. When multiplied, the area is expressed in square centimeters. If length and width are made up of mixed units, they must be converted to one or the other before the area is calculated.

To show how this works, let's measure the width and length of a floor. Suppose we measure 4 meters. We have not quite reached the other wall, and the meter measure is too long to make it another time. So we use the centimeter divisions on the meter stick, and note that the remaining distance is 35 centimeters. So the width of the room is 4 meters, 35 centimeters. Then we measure the length of the room and find it is 7 meters, 82 centimeters.

Before anything more can be done, centimeters must be converted to meters. Keeping in mind that a centimeter is $\frac{1}{100}$ of a meter, the decimal must be moved two places to convert centimeters into meters. That is, 35 centimeters equal 0.35 meter. And 82 centimeters equal 0.82 meter. Now these figures in meters can be added to the whole numbers of meters. Thus the width of the room is $4 + 0.35$ or 4.35 meters. And the length is $7 + 0.82$ or 7.82 meters. To find the area, multiply 4.35 meters by 7.82 meters for an area of 34.02 square meters, rounded off to the second decimal place.

RELATION OF METRIC AREAS

Metric areas can vary from a small one, such as a square centimeter, up to a very large one, such as a square kilometer. Here are how these areas are related to each other:

1 square centimeter is a square 1 cm by 1 cm.
1 square meter is a square 1 m by 1 m.
1 square meter = 10,000 square centimeters
1 are is a square 10 m by 10 m
1 are = 100 square meters
1 hectare is a square 100 m by 100 m
1 hectare = 100 ares = 10,000 square meters
1 square kilometer is a square 1,000 m by 1,000 m
1 square kilometer = 100 hectares = 10,000 ares

Don't try to memorize these figures. They are given only for reference, in case you need them sometime.

Here are some ways of calculating metric areas:

1. Width in centimeters × length in centimeters = area in square centimeters.

2. Width in meters × length in meters = area in square meters.

3. Width in kilometers × length in kilometers = area in square kilometers.

4. Find area in square meters and move decimal two places to the left to obtain area in ares.

5. Find area in square meters and move decimal four places to the left to obtain area in hectares.

6. To convert ares to hectares or the reverse, move decimal two places in the proper direction, keeping in mind that a hectare is 100 times as big an area as an are.

7. To convert hectares to square kilometers, or the reverse, move decimal two places in the proper direction, keeping in mind that a square kilometer is 100 times as big an area as a hectare.

KEY AREA CONVERSIONS

If you *must* make a conversion to English units, here are some key conversions to remember:

A square inch is about $6\frac{1}{2}$ times as big as a square centimeter.

A square yard is just a little smaller than a square meter.

A hectare is about $2\frac{1}{2}$ acres.

There are about 2.6 square kilometers in 1 square mile.

PROBLEMS

Since you were not asked to memorize the conversion factors between the various metric units of area, you may refer to the preceding pages to work these problems.

1. Convert the following:
   a. 1.75 square meters to square centimeters
   b. 2,400 square meters to ares
   c. 450 ares to hectares
   d. 450 ares to square kilometers
   e. 2.45 square kilometers to hectares

f. 4.85 hectares to ares

g. 20 ares to square meters

2. If a farm measures 500 meters × 1,500 meters, what is its area in (a) square kilometers, (b) hectares, (c) ares?

3. Multiply 4 meters, 25 centimeters by 3 meters, 45 centimeters to get the area in (a) square meters, (b) square centimeters.

4. If a living room is 10 feet × 15 feet, how much will it cost to carpet it, if carpeting costs $9 a square meter? (Hint: convert 10 feet and 15 feet to meters, then find the area in square meters. Use 1 inch equals 2.5 centimeters. You should know the number of inches in a foot and the number of centimeters in a meter.)

5. A wall 1 meter 50 centimeters long and 1 meter 10 centimeters high is to be covered with tile. Each piece of tile is 11 centimeters × 11 centimeters. What is the minimum number of pieces of tile needed?

6. An area contains 50 hectares. About how many acres is this?

7. A room is to be wall-papered. It is 4 meters 25 centimeters wide, 5 meters 60 centimeters long, and 2 meters 80 centimeters high. In the room there are two doors that are 1 meter wide and 2 meters 30 centimeters high. Also one window that is 1

meter 50 centimeters wide and 2 meters high. How many square meters of wallpaper will be needed? (Remember that a room has four walls and that no wallpaper is needed for the doors and the window.)

8. If a dress pattern calls for 3.5 yards of material 45 inches wide, how many meters will be required of material 110 centimeters wide? (You will need the conversion factor: 10.76 square feet = 1 square meter.) The area of material needed is exactly the same whether it is measured in English or in metric units. As the last step in the problem:

(length in meters) (width in meters) = (area in square meters)

or

$$(\text{length in meters}) = \frac{(area\ in\ square\ meters)}{(\text{width in meters})}$$

9. Use a centimeter ruler to get the length and width of the front cover of this book to the nearest tenth of a centimeter. Then calculate the area of the cover in square centimeters.

Answers to problems begin on page 104.

4

Volume

Volume is the three-dimensional space occupied by a material. This material can be anything—water, metal, plastic, air, gasoline, etc. Areas are two-dimensional. That is, to find the area of a rectangle or a square, you multiply the length of one side by the length of the other side. Volume goes one step further. It is three-dimensional. To obtain the volume inside a regularly shaped box, multiply the length by the width by the height.

How volumes are measured depends on the size of the volume. Fairly small volumes are measured in cubic centimeters or in milliliters. Larger volumes are measured in liters. Still larger volumes may be expressed in cubic meters.

Let's talk about liters first. A liter is a cube 10 centimeters wide, 10 centimeters long, and 10 centimeters high. That is, a liter contains $10 \times 10 \times 10 = 1,000$ cubic centimeters (cm³).

To help visualize the size of a liter, make a liter cube by drawing, cutting out, folding, and gluing a pattern such as the one in Figure 5. Use a large sheet of heavy paper or cardboard. Use a metric ruler to obtain the correct lengths as marked on Figure 5. Make three flaps large enough so they can be folded over and glued to hold the cube together.

The finished cube should look like the cube shown on the right in Figure 5.

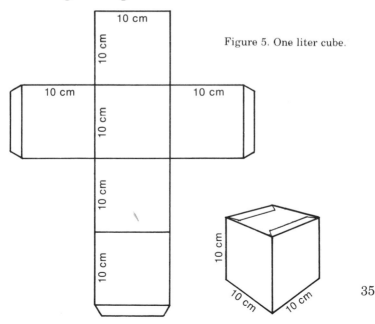

Figure 5. One liter cube.

For practice, calculate the total surface of the cube in square centimeters. Remember that a box has six sides.

The volume of one liter can occupy a shape other than a cube. It can even be an irregular shape, such as the inside of a pitcher. To illustrate this, you may want to make another box, using the dimensions shown in Figure 6.

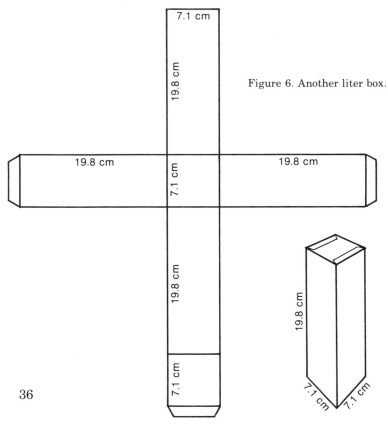

Figure 6. Another liter box.

When the pattern in Figure 6 is put together, it makes a taller, narrower box, but its volume is still 1,000 cubic centimeters or one liter. Now calculate the total surface area of this tall, narrow box, finding the answer in square centimeters to one decimal place.

You should have gotten a surface area of 600 square centimeters for the one liter cube, and 663.1 square centimeters for the tall, narrow box. Are you surprised that the surface area of the two boxes is not the same, even though the volume inside of each box is the same?

A ball, or sphere, has the smallest possible surface area for a particular volume. A sphere with a volume of 1,000 cubic centimeters, or one liter, has a diameter of 12.4 centimeters, and a surface area of only 482 square centimeters. This is considerably less than the figures given above for the two boxes.

A liter is about 5% bigger than a quart. If milk is packaged in a liter container, you will be getting a "big quart." Gasoline can be measured by the liter. Four liters of gasoline are a little more than a gallon. So if you use 8 gallons of gasoline to fill your car tank, you will need a little less than 8×4, or 32 liters of gasoline.

MILLILITERS AND CUBIC CENTIMETERS

Since a liter contains 1,000 cubic centimeters, it is evident that one cubic centimeter is $\frac{1}{1000}$ of a liter.

There is also a small unit of volume called a milliliter. By now, you should know that a milliliter must be $\frac{1}{1000}$ of a liter.

You may take a look at the last paragraph and say, "Hey, wait a minute. If a cubic centimeter is $\frac{1}{1000}$ of a liter, and a milliliter is also $\frac{1}{1000}$ of a liter, then a cubic centimeter and a milliliter must be exactly the same volume." You are right! One cubic centimeter equals one milliliter. Both measures are $\frac{1}{1000}$ of a liter. It doesn't matter whether you say a volume is 475 cubic centimeters or 475 milliliters.

Some people always ask questions, and I can hear one now: "Why have two units that are exactly the same?" Here is a simple explanation: A cubic centimeter is a cube 1 centimeter by 1 centimeter by 1 centimeter. It is based on metric units of length— namely centimeters. A milliliter is defined as $\frac{1}{1000}$ of a standard volume called a liter. So a milliliter is based on a metric unit of volume—namely the liter. The metric system was designed so these two measures would be exactly equal. You'll have to admit that those Frenchmen were really clever!

Note: A cubic centimeter and a milliliter are really not *exactly* equal, but the difference is so small that they are equal for all practical purposes.

To find the volume of a regularly shaped container, such as a box, multiply the width in centimeters by the length in centimeters, by the height in centimeters. If a box is 20 centimeters wide, 30 centimeters long, and 25 centimeters high, its volume is $20 \times 30 \times 25 =$ 15,000 cubic centimeters (cm³).

Note: The answer for $4 \times 4 \times 4$ can be expressed as 4^3. Similarly, if a length in cm is multiplied by a width in cm, then by a height in cm, the answer is cm³. So this is the abbreviation for "cubic centimeters."

Small items often show the volume in fluid ounces. An equal volume in cubic centimeters (or milliliters) will be close to 30 times as great a number. Thus a 4-fluid-ounce bottle holds 4×30, or 120 cubic centimeters. An 8-fluid-ounce bottle holds 8×30, or 240 cubic centimeters. An average-sized cup holds about 200 cm³, and an average-sized drinking glass holds about 300 cm³.

A word of warning: The English system is confusing because it has fluid ounces and weight ounces. Be sure to look on the bottle to see which one is being used. The conversion discussed above is only for fluid ounces. Confusion between the kinds of ounces will disappear when we use only the metric system.

CUBIC METERS

A cubic meter (m³) is a cube that is 1 meter wide, 1 meter long, and 1 meter high. A cubic meter contains 1,000 liters. It is used to measure large volumes. For example, the rate flow of pumps is commonly expressed in cubic meters per hour. Or the volume of a room can be calculated in cubic meters. Suppose a room is 4.75 meters wide, 6.25 meters long, and 2.52 meters high. Its volume is $4.75 \times 6.25 \times 2.52 = 74.82$ cubic meters, to two decimal places.

IRREGULAR VOLUMES

Often a three-dimensional space is not regular in shape. A pitcher or a fancy bottle are good examples. It is possible to find the volume of irregular objects by using a graduate marked with metric measurements of volume. (See Figure 7.) Fill the irregularly shaped container with water, and pour the water into the graduate. Then all you have to do is to read the volume in cubic centimeters at the surface of the water in the graduate. For

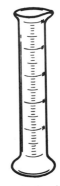

Figure 7. Graduate.

small volumes, measuring cups graduated in metric units are useful. Metric units will replace the fluid ounces on the measuring cups used in the kitchen now.

REVIEWING VOLUMES

In review, here are the metric units of volume:

1 liter = 1,000 milliliters
1 liter = 1,000 cubic centimeters

1 milliliter = 1 cubic centimeter
1 cubic meter = 1,000 liters
A cubic centimeter is a cube 1 centimeter × 1 centimeter × 1 centimeter.
A cubic meter is a cube 1 meter × 1 meter × 1 meter.

Here are a few key conversions between metric and English units of volume:

1 fluid ounce = about 30 cubic centimeters or about 30 milliliters
1 liter is about 5% bigger than a quart
4 liters are a little more than a gallon
A cubic meter is a little bigger than a cubic yard.

PROBLEMS

1. Each side of a cube is 6 centimeters. What is its volume in cubic centimeters?

2. A corrugated container is 44 centimeters long, 32 centimeters wide, and 28 centimeters high. What is its volume (a) in cubic centimeters, (b) in liters?

3. Calculate the total surface area of the corrugated container of problem 2. Get the answer (a) in square centimeters, (b) in square meters. (Remember that such a container has *six* sides.)

4. If a certain volume is 1,500 cubic centimeters, how many milliliters is it?

5. A small bottle holds 150 cubic centimeters. How many fluid ounces is this?

6. If a large container holds 10,500 liters, what is its volume in cubic meters?

7. Ten liters is about how many quarts?

8. If you usually need 10 gallons of gasoline to fill your car tank, about how many liters will you need?

Answers to problems begin on page 107.

5

Connecting Volume
and Weight

Often it is desired to know the weight of a certain volume of a given material, or what volume will be occupied by a certain weight of that material. It seems that nothing was left out when the metric system was devised, because there is a very simple relationship between metric weight and volume. If a volume of one liter is filled with water that has a temperature of 4° Celsius (C), the weight of the water is exactly one kilogram. The temperature is mentioned because water expands slightly as it is heated. Except for very accurate scientific work, you need not be concerned about this.

Since one cubic centimeter (or one milliliter) is $\frac{1}{1000}$ of a liter, and since one gram is also $\frac{1}{1000}$ of a liter, it is also true that 1 cubic centimeter (or 1 milliliter) of water 4°C weighs 1 gram.

This makes it very easy to calculate the weight of any volume of water. Thus 325 cubic centimeters of water weigh 325 grams. Or 2.47 liters of water weigh 2.47 kilograms. What could be simpler?

DENSITY

So you say "That's fine if the material is water. But what do I do if the volume is filled with aluminum, copper, iron, alcohol, or glass? Then how much does it weigh?"

To determine this, all you need to know is the *density* of a material and then make a simple multiplication.

Density is a measure of how much heavier or lighter a material is than water. For example, copper has a density of 8.92. This means that copper is 8.92 times as heavy as an equal volume of water. If 200 cubic centimeters of water weigh 200 grams, then 200 cubic centimeters of copper weigh 200 × 8.92 or 1,784 grams.

In the metric system density is the weight in grams of one cubic centimeter (or one milliliter) of a material. Or density is the weight in kilograms of one liter of a material.

Here are the densities of some common materials:

Materials	Density
	*(g/cm³ or kg/1)
Aluminum	2.70
Brick	1.4−1.6
Butter	0.94−0.95
Cast iron	7.25
Copper	8.92
Gasoline	0.66−0.69
Gold	19.3
Lead	11.34
Magnesium	1.74
Mercury	13.55
Milk	1.028−1.035
Nickel	8.90
Sea water	1.025
Silver	10.5
Steel	7.6−7.8
Window glass	2.4−2.6
Zinc	7.14

*Grams per cubic centimeter, or kilograms per liter.

FINDING WEIGHT FROM VOLUME

To find the weight of any known volume, you multiply the volume by the density of the material. For
example, the density of aluminum is 2.70 grams per
cubic centimeter. So the weight of 375 cubic centi-

meters of aluminum is $375 \times 2.70 = 1,012.5$ grams. The density of milk is 1.03 grams per cubic centimeter. So the weight of 200 cubic centimeters of milk is $200 \times 1.03 = 206$ grams. The same procedure can be used to get the weight in grams of a known volume of any material—whether it is light or heavy, or whether it is a gas, liquid, or solid.

Although density is usually expressed as grams per cubic centimeter, it can also be expressed as kilograms per liter. That is, aluminum has a density of 2.70 kilograms per liter. Thus the weight of 150 liters of aluminum is $150 \times 2.70 = 405$ kilograms.

Here are two simple formulas for calculating the weight of any known volume of a material:

1. (volume in cubic centimeters) (density of material in grams per cubic centimeter) = weight in grams
2. (volume in liters) (density in kilograms per liter) = weight in kilograms

Remember that density in kilograms per liter is exactly the same number as density in grams per cubic centimeter.

The same method is used to find the weight of any volume of a material that is lighter than water. For example, butter has a density of about 0.94 gram per cubic centimeter. So the weight of 850 cubic centimeters of butter is $850 \times 0.94 = 799$ grams. The density of water at 40°C is 0.99224 gram per cubic centimeter. So, 350 cubic centimeters of 40°C water weigh $350 \times 0.99224 = 347.284$ grams. This compares with 350.000 grams for 350 cubic centimeters of water at 4°C.

Now let's go in the reverse direction. What if you want to know the volume that a material occupies if you know its weight? Here are the two formulas that can be used:

1. $\dfrac{\text{weight in grams}}{\text{density}} = $ volume in cubic centimeters

2. $\dfrac{\text{weight in kilograms}}{\text{density}} = $ volume in liters

If you have 480 grams of cast iron, what volume does it occupy? The density of cast iron is 7.25 grams per cubic centimeter. So we have:

$$\frac{480}{7.25} = 66.21 \text{ cubic centimeters}$$

If you have 2.45 kilograms of magnesium, what volume does it occupy? The density of magnesium is 1.74. So we have:

$$\frac{2.45}{1.74} = 1.41 \text{ liters}$$

The density figures given in the above examples show that metals vary greatly in how heavy they are, for equal volumes. Magnesium is the lightest of the common metals, with a density of 1.74. Aluminum is a little heavier, with a density of 2.70. Then, in order, come zinc (7.14), cast iron (7.25), copper (8.92), lead (11.34), mercury (13.55), and gold (19.3). If you ever have a chance to lift a gold brick, you will find that for its size it is very heavy. For equal volumes, gold weighs 19.3 times as much as water.

PROBLEMS

1. If a container with a volume of 450 cubic centimeters is filled with water, how much will the water weigh (a) in grams, (b) in kilograms?

2. Another container weighs 100 grams empty. When filled with water, its total weight is 475 grams. What is the volume of the container in (a) cubic centimeters, (b) liters?

3. A pitcher with a volume of 2.47 liters is filled with water. How much does the water weigh in (a) kilograms, (b) grams?

4. An aluminum bar is 5 centimeters wide, 3 centimeters thick, and 1 meter 60 centimeters long. The density of aluminum is 2.70 grams per cubic centimeter. What is the weight of the bar in (a) grams, (b) kilograms?

5. A piece of copper weighs 12 kilograms 340 grams. What is its volume in (a) liters, (b) cubic centimeters? (The density of copper is 8.92 grams per cubic centimeter = 8.92 kilograms per liter.)

6. An aquarium is 30 centimeters wide, 40 centimeters long, and 20 centimeters high. When filled with water, what is the weight of the water in (a) grams, (b) kilograms?

7. A swimming pool is 7 meters wide, 14 meters long and has an average depth of 1.4 meters from the bottom to the surface of the water. What is the weight of the water in the pool in (a) kilograms, (b) metric tons, (c) U.S. short tons (approximate)?

Answers to problems begin on page 108.

6

How Hot or Cold Is It?

When things change, they *really* change! You must learn not only the metric units of weight, length, area and volume, but you must also learn how to interpret a different kind of thermometer. I learned to call it a centigrade thermometer, but it is now called a Celsius thermometer. Keep in mind, however, that the two names refer to the same temperature scale. From here on, we'll use the name *Celsius*.

With a Fahrenheit thermometer (the one you have been looking at all these years), water freezes at 32° and boils at 212°. There are 180 degrees between the freezing and boiling points of water on a Fahrenheit thermometer. On the Celsius thermometer, 0° is the freezing point and 100° is the boiling point of water—a nice even amount of 100 degrees lies between the freezing and boiling points.

How are you going to know how hot or cold it is when you read a Celsius thermometer? Right away you have two reference points. If the thermometer reads 0°, it is just cold enough for water to start freezing. If it reads 100°, then it is hot enough to boil water. (Most household Celsius thermometers don't read higher than 50°.)

I like to remember another figure. A Celsius reading of 20° is the same as 68°F. So 20°C is a comfortable temperature—not hot and not cold.

In Figure 8 Celsius and Fahrenheit thermometers are compared. You will notice that −10°C is +14°F; this is because Celsius readings are negative as soon

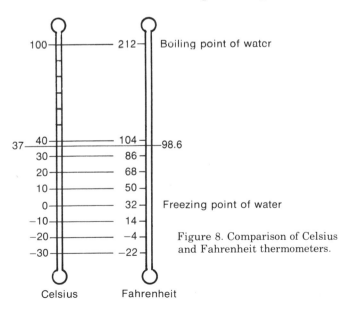

Figure 8. Comparison of Celsius and Fahrenheit thermometers.

Celsius   Fahrenheit

as the temperature falls below the freezing point of water.

In Figure 8, I have included 37° Celsius for a particular reason. Normal body temperature is 98.6°F, and this is equal to 37°C. So if you read someone's temperature and it is 39°C, you will know that person has some fever.

On a terribly hot day, a Fahrenheit thermometer might read 104°, while a Celsius thermometer will read only 40°. I'm sorry about it, but it is exactly as hot at 40°C as it is at 104°F.

CONVERTING CELSIUS READINGS TO FAHRENHEIT

For a while it may be difficult for you to interpret Celsius temperatures. Of course you can carry a conversion chart around with you, but I don't like to use any conversion chart if I can avoid it. Instead, I suggest that you memorize the following:

$$0°C = 32°F$$
$$10°C = 50°F$$
$$20°C = 68°F$$
$$30°C = 86°F$$
$$100°C = 212°F$$

For temperatures in-between, you will note that for 10 degrees Celsius, you have 18 degrees Fahrenheit. So the exact halfway value between 10 and 20°C is equal to nine degrees Fahrenheit. Thus 15°C is 50 + 9, or 59°F and 25°C is 68 + 9, or 77°F. This you can do in your head, if you have memorized the conversions for 10°C and 20°C.

If you want to get a conversion that is closer than 5 degree intervals, use 2 degrees Fahrenheit for every 1 degree Celsius. This isn't exactly true, but it is accurate enough. Here are some examples:

$$21°C = 68 + 2 = 70°F$$
$$18°C = 68 - 4 = 64°F$$
$$32°C = 86 + 4 = 90°F$$

This method will work for Celsius readings that are one degree or two degrees from 10, 20 or 30°C. For a conversion of readings such as 3°C, 14°C, or 27°C to the corresponding Fahrenheit readings, it will be necessary to memorize the following:

$$5°C = 41°F$$
$$15°C = 59°F$$
$$25°C = 77°F$$
$$35°C = 95°F$$

Now conversions of temperatures near these values can be carried out in your head. Here are some examples:

$$3°C = 41 - 4 = 37°F$$
$$14°C = 59 - 2 = 57°F$$
$$27°C = 77 + 4 = 81°F$$

As with the other metric units, the idea is to stop making mental conversions just as soon as you can. Probably the best way to make the change is to have

Celsius thermometers indoors and outdoors, and to look at them frequently. Then compare the reading with how hot or cold you feel, and try to mate the two. Good luck!

Don't attempt the problems below until you have memorized the key conversions of 0°C, 10°C, 20°C, and 30°C.

1. Without looking at the conversion chart or the material in this chapter, give the approximate Fahrenheit readings for:

29°C	9°C
22°C	2°C
19°C	−1°C
12°C	

2. From the key conversions, give *accurate* Fahrenheit readings for:

35°C	5°C
25°C	−5°C
15°C	

3. If Mary's temperature is 38°C, does she have a fever? Why?

Answers to problems begin on page 109.

7

Learning to "Think Metric"

All of us in the United States are familiar with the English system. We know about how long a foot is, or a yard. We have at least an approximate idea of how heavy a pound is. We know the size of a quart, such as a quart of milk.

What is needed now is to gain the same familiarity with some of the common metric units, so you can "Think Metric." This is not as difficult as it sounds if you follow some simple rules and methods. Here are a few:

Measure many things with a metric tape or meter stick. Then remember that a table is so many centimeters wide, or that a door is so many meters high. *Don't* convert these figures to the English system.

Make a liter container so you will know how big its volume is.

Watch for items in the supermarket that are labeled with the weight in grams. Then lift that object and begin to see how heavy it feels in relationship to its weight in grams.

Be sure to keep in mind that the metric system is very simple. All units are related to each other by factors of 10, 100, 1,000, or 10,000. This means that conversions from one to the other can often be made mentally, without the need for pencil and paper. For example, 350 grams are 0.35 kilogram. Or 450 centimeters are 4.5 meters. Or 2,500 milliliters are 2.5 liters. This simplicity is what makes the metric system such a fine one. Of course, you must know that there are 1,000 grams in a kilogram, 100 centimeters in a meter, and 1,000 milliliters in a liter.

It is also important to keep in mind that the calculation of areas and volumes in the metric system is no different than it is in the English system. You merely use metric lengths instead of English lengths.

When you start working with the metric system it is necessary to make at least rough conversions to the English equivalents, so you will have some idea of the magnitude of the metric measurements. You will notice that not a single conversion table has been included in Part I of this book. Instead, a small number of "key" conversions have been given. Memorizing these conversions will help in figuring the approximate size of certain metric measurements. Then you should make every effort to stop doing this just as soon as possible.

The learning of anything new requires some effort. But I am certain that once you have learned to "Think Metric" you will say, "How did we ever endure that English system for such a long time?"

PART TWO

8

Metric Prefixes

For many people, it will be necessary to learn only the material in Part One in order to get along in a "metric world." The information included in Part Two is for those people who want to know more, or who have special calculations to make.

The use of metric prefixes is one example. In Chapter 1, the prefixes centi-, milli-, and kilo- were presented. These are the ones used most commonly. However, there are a number of other standard prefixes that range from very large numbers to extremely small numbers. There is also a symbol, or abbreviation, assigned to each prefix. Here is the list:

Prefixes for big numbers:			*Symbol*
tera-	1,000,000,000,000.	or 10^{12}	T
	(a million million)		
giga-	1,000,000,000.	or 10^9	G
	(1 billion)		

mega-	1,000,000.	or 10^6	M
kilo-	1,000.	or 10^3	k
hecto-	100.	or 10^2	h
deka-	10.	or 10^1	da

Prefixes for small numbers:

deci-	0.1	or 10^{-1}	d
centi-	0.01	or 10^{-2}	c
milli-	0.001	or 10^{-3}	m
micro-	0.000001	or 10^{-6}	μ
nano-	0.000000001	or 10^{-9}	n
pico-	0.000000000001 or 10^{-12}		p

Here are some examples of prefixes of big numbers:

A dekameter is 10 meters.
A hectare is 100 ares.
A kilowatt is 1,000 watts.

A megacycle is 1,000,000 cycles.

Here are some examples of prefixes of small numbers:

A decimeter is $\frac{1}{10}$ of a meter.
A micrometer is $\frac{1}{1000000}$ of a meter. It is also $\frac{1}{10000}$ of a centimeter, or $\frac{1}{1000}$ of a millimeter. It is more commonly called a micron. (This term should not be confused with a measuring instrument called a micrometer.)
A centigram is $\frac{1}{100}$ of a gram.

A nanometer is $\frac{1}{1000000000}$ (one-billionth) of a meter.

Wavelengths of visible light are often expressed in nanometers.

They vary from about 400 nanometers (in the deep blue) to about 700 nanometers (in the deep red).

A nanosecond is $\frac{1}{1000000000}$ (one-billionth) of a second. Computer processing speeds are often measured in nanoseconds.

The use of these standard prefixes adds to the simplicity of the metric system. It doesn't matter what term follows the prefix. For instance, if you see the word "microsecond" you can be certain that what is meant is one-millionth of a second.

9

Metric-Metric Conversion

If you have memorized the metric prefixes (see Chapter 8), you will know many of the conversions listed below without looking at them. "Deci-" means $\frac{1}{10}$, so a deciliter must be 0.1 liter, and a decigram must be 0.1 gram. The prefix "hecto-" means 100, so a hectoliter is 100 liters, and a hectogram is 100 grams. The prefix "hecto-" is shortened to "hect-" to give "hectare," and a hectare is 100 ares.

In some conversions, the power of 10 is included. The power number indicates the number of places the decimal must be moved to make the conversion. When the power number is positive, the decimal point is moved to the right. For example, if 1 square kilometer equals 10^4 ares, 5.73 square kilometers equals 57,300 ares. The decimal was moved to the right four places.

When a number like 10^{-3} is involved, then the decimal must be moved to the left three places. For example, 1 milligram equals 10^{-3} gram. So 568 milligrams equals 0.568 gram. In this case the decimal was moved to the left three places.

Suppose you want to convert grams to kilograms. If you look under "gram" in the table you will find "1 gram = 0.001 kilogram." If you don't like to use a small number like 0.001, then look under "kilogram," and you will find "1 kilogram = 1,000 grams."

Most metric-metric conversions can be made merely by moving the decimal point the correct number of places. The table below lists many metric-metric conversions:

METRIC-METRIC CONVERSIONS

1 are (a) = 100 square meters (m²)

1 are (a) = 0.01 hectare (ha)

1 are (a) = 0.0001 square kilometer (km²) (10^{-4})

1 atmosphere = 76.0 centimeters (cm) of mercury

= 760 millimeters (mm) of mercury

1 carat (metric) = 200 milligrams (mg)

1 centimeter (cm) = 0.1 decimeter (dm)

1 centimeter (cm) = 10 millimeters (mm)

1 centimeter (cm) = 0.01 meter (m)

1 centimeter (cm) = 10,000 microns (μ) (10^4)

1 centimeter (cm) = 10^7 millimicrons (mμ)

1 centimeter/second (cm/sec) = 0.036 kilometer/hour (km/hr)

1 cubic centimeter (cm³) = 1,000 cubic millimeters (mm³)

Metric-Metric Conversion   65

1 cubic centimeter (cm³) = 0.001 liter (l)
1 cubic centimeter (cm³) = 1 milliliter (ml)
1 cubic centimeter (cm³) = 0.001 cubic decimeter (dm³)
1 cubic centimeter (cm³) of water at 4°C = 1 gram (g)
1 cubic decimeter (dm³) = 1,000 cubic centimeters (cm³)
= 1,000 milliliters (ml)
1 cubic decimeter (dm³) = 1 liter (l)
1 cubic meter (m³) = 1,000 liters (l)
1 cubic millimeter (mm³) = 0.001 cubic centimeter (cm³)
1 decigram (dg) = 0.1 gram (g)
1 decimeter (dm) = 0.1 meter (m)
1 decimeter (dm) = 10 centimeters (cm)
1 dekagram (dag) = 10 grams (g)
1 dekaliter (dal) = 10 liters (l)
1 dekameter (dam) = 10 meters (m)
1 gram (g) = 10 decigrams (dg)
1 gram (g) = 0.1 dekagram (dag)
1 gram (g) = 0.01 hectogram (hg)
1 gram (g) = 1,000 milligrams (mg)
1 gram (g) = 0.001 kilogram (kg)
1 gram/cubic centimeter (g/cm³) = 1 kilogram/liter (kg/l)
1 gram (g) of water at 4°C = 1 cubic centimeter (cm³)
= 1 milliliter (ml)
1 gram/liter (g/l) = 1,000 parts/million (ppm)
1 hectare (ha) = 100 ares (a)
1 hectare (ha) = 10,000 square meters (m²) (10^4)
1 hectare (ha) = 0.01 square kilometer (km²)
1 hectogram (hg) = 100 grams (g)
1 hectoliter (hl) = 100 liters (l)
1 hectometer (hm) = 100 meters (m)

1 horsepower (metric) = 75 kilogram-meters/second (kg-m/sec)

1 kilogram (kg) = 1,000 grams (g)

1 kilogram (kg) = 0.001 metric ton

1 kilogram/liter (kg/l) = 1 gram/cubic centimeter (g/cm³)

= 1 gram/milliliter (g/ml)

1 kilogram (kg) of water at 4°C = 1 liter (l)

1 kiloliter (kl) = 1,000 liters (l)

1 kilometer (km) = 1,000 meters (m)

1 kilometer/hour (km/hr) — 27.78 centimeters/second (cm/sec)

1 kilowatt (kw) = 1,000 watts (w)

1 liter (l) = 0.01 hectoliter (hl)

1 liter (l) = 1,000 milliliters (ml)

1 liter (l) = 1,000 cubic centimeters (cm³)

1 liter (l) — 0.001 cubic meter (m³)

1 liter (l) = 1 cubic decimeter (dm³)

1 liter (l) = 0.1 dekaliter (dal)

1 liter (l) = 0.001 kiloliter (kl) (10⁻³)

1 liter (l) of water at 4°C = 1,000 grams (g)

= 1 kilogram (kg)

1 meter (m) = 0.1 dekameter (dam)

1 meter (m) = 0.01 hectometer (hm)

1 meter (m) = 0.001 kilometer (km) (10⁻³)

1 meter (m) = 100 centimeters (cm)

1 meter (m) = 10 decimeters (dm)

1 meter (m) = 1,000 millimeters (mm)

1 meter/second (m/sec) = 3.6 kilometers/hour (km/hr)

1 milligram (mg) = 0.001 gram (g) (10⁻³)

1 milligram/liter (mg/l) = 1 part per million (ppm)

1 milliliter (ml) = 0.001 cubic decimeter (dm³)

1 milliliter (ml) = 1 cubic centimeter (cm³)

1 milliliter (ml) = 0.001 liter (l)

1 milliliter (ml) of water at 4°C = 1 gram (g)

1 millimeter (mm) = 0.001 meter (m) (10^{-3})

1 millimeter (mm) = 0.1 centimeter (cm)

1 square centimeter (cm²) = 0.0001 square meter (m²) (10^{-4})

1 square centimeter (cm²) = 100 square millimeters (mm²)

1 square centimeter (cm²) = 0.01 square decimeter (dm²)

1 square decimeter (dm²) = 100 square centimeters (cm²)

1 square decimeter (dm²) = 0.01 square meter (m²)

1 square kilometer (km²) = 10,000 ares (a) (10^{4})

1 square kilometer (km²) = 100 hectares (ha)

1 square kilometer (km²) = 1,000,000 square meters (m²) (10^{6})

1 square meter (m²) = 10,000 square centimeters (cm²) (10^{4})

1 square meter (m²) = 100 square decimeters (dm²)

1 square meter (m²) = 0.0001 hectare (ha) (10^{-4})

1 square meter (m²) = 0.01 are (a)

1 square meter (m²) = 10^{-6} square kilometer (km²)

= 0.000,001 (km²)

1 square millimeter (mm²) = 0.01 square centimeter (cm²)

1 ton (metric) = 1,000 kilograms (kg)

10

Converting Celsius and Fahrenheit Temperature Readings

Most people will be able to interpret Celsius thermometer readings closely enough by following the suggestions given in Chapter 6. The conversion chart below is *only* for those who want an accurate conversion reading.

CELSIUS-FAHRENHEIT CONVERSIONS

°C	°F	°C	°F	°C	°F
−40	−40	−26	−14.8	−16	3.2
−38	−36.4	−24	−11.2	−15	5
−36	−32.8	−22	− 7.6	−14	6.8
−34	−29.2	−20	− 4	−13	8.6
−32	−25.6	−19	− 2.2	−12	10.4
−30	−22	−18	− 0.4	−11	12.2
−28	−18.4	−17	1.4	−10	14

°C	°F	°C	°F	°C	°F
− 9	15.8	14	57.2	38	100.4
− 8	17.6	15	59	39	102.2
− 7	19.4	16	60.8	40	104
− 6	21.2	17	62.6	42	107.6
− 5	23	18	64.4	44	111.2
− 4	24.8	19	66.2	46	114.8
− 3	26.6	20	68	48	118.4
− 2	28.4	21	69.8	50	122
− 1	30.2	22	71.6	55	131
		23	73.4	60	140
0	32	24	75.2	65	149
1	33.8	25	77	70	158
2	35.6	26	78.8	75	167
3	37.4	27	80.6	80	176
4	39.2	28	82.4	85	185
5	41	29	84.2	90	194
6	42.8	30	86	95	203
7	44.6	31	87.8		
8	46.4	32	89.6	100	212
9	48.2	33	91.4		
10	50	34	93.2	125	257
11	51.8	35	95	150	302
12	53.6	36	96.8	175	347
13	55.4	37	98.6	200	392

One Celsius degree is equal to 1.8 (or $\frac{9}{5}$) Fahrenheit degrees. This leads to the general equation for converting any Celsius reading to Fahrenheit. It is:

$$°F = 32 + (°C)(1.8)$$

Use this formula to convert 200°C to °F:

$$°F = 32 + (200)(1.8) = 32 + 360 = 392°F$$

If a reading is negative, then you must take the negative sign into account. Thus to convert −30°C to °F:

$$°F = 32 + (−30)(1.8) = 32 − 54 = −22°F.$$

Another formula can be used to convert any Fahrenheit temperature to Celsius. It is:

$$°C = \tfrac{5}{9}(°F. − 32)$$

For example, to convert 68°F. to °C:

$$°C = \tfrac{5}{9}(68 − 32) = \tfrac{5}{9}(36) = 20°C.$$

If a conversion table, as the one above, is available, there is no need for these formulas. And when you learn to relate a Celsius reading to how hot or cold it is, you can dispense with all conversions to Fahrenheit.

11

Calculating Areas and Volumes

In Chapters 3 and 4, only simple shapes were included in the discussions of area and volume. When other shapes are involved, only the formula for the calculation of the area or volume is required. These formulas are exactly the same as the ones used in the English system—you merely use metric measurements instead of English.

Measurements must be consistent; that is, if one side of an area is measured in centimeters, the other side must also be in centimeters, and the area is calculated as so many square centimeters.

Here are formulas for the calculation of some common areas and volumes:

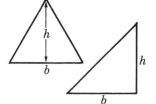

Triangle

Area = $\frac{1}{2}\, bh$

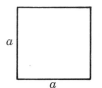

Square

Area = $a \times a$

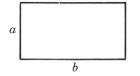

Rectangle

Area = $a \times b$

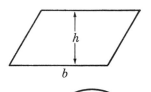

Parallelogram

Area = $b \times h$

Circle

Area = $\pi\, r^2$ where r is the radius of the circle, and π is 3.1416 (or 3.14 for many calculations)

Calculating Areas and Volumes   73

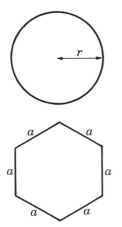

Sphere

Area = $4 \pi r^2$ where r is the radius of the sphere

Regular hexagon

Area = $2.598\ a^2$ where a is the length of each of the six sides

VOLUMES

Cube

Volume = a^3 $(a \times a \times a)$

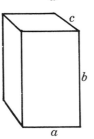

Rectangular box

Volume = $a \times b \times c$

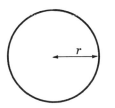

Sphere

Volume $= \dfrac{4\ \pi\ r^3}{3}$ where r is the radius of the sphere

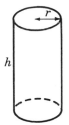

Cylinder

Volume $= \pi\ r^2 h$ where r is the radius of the base circle and h is the height

Cone

Volume $= \dfrac{\pi\ r^2 h}{3}$ where r is the radius of the base circle, and h is the vertical height

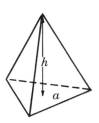

Pyramid

Volume $= \dfrac{ah}{3}$ where a is the *area* of the base triangle, and h is the vertical height

As an example of the use of these formulas, let's calculate the area of a circle that has a radius, r, of 0.4 meter:

$$\text{area} = \pi r^2 = (3.1416)(0.4^2) = (3.1416)(0.16)$$
$$= 0.503 \text{ square meter}$$

The volume of a sphere with a radius of 8 cm is calculated as follows:

$$\text{volume} = \frac{4\pi r^3}{3} = \frac{4 \times 3.1416 \times 8^3}{3} = \frac{4 \times 3.1416 \times 512}{3}$$
$$= 2{,}143.74 \text{ cm}^3$$

PROBLEMS

1. A triangle has a base of 40 centimeters and a height of 25 centimeters. What is its area? Give units of area also.

2. If a parallelogram has a base of 30 centimeters and a vertical height of 15 centimeters, what is its area? Give units of area also.

3. What is the area of a circle with a radius of 8 centimeters? (Use 3.14 for π.)

4. If a sphere has a radius of 8 centimeters, what is the area of its surface? (Use 3.14 for π.)

5. A cylinder is 1.1 meters long and the radius of its base circle is 7 centimeters. What is its volume (a) in cubic centimeters, (b) in cubic meters?

Answers begin on page 110.

12

English-Metric Conversion

The conversion factors listed in this chapter are only for use until the United States is 100 percent metric. Accurate conversion factors are given. This doesn't mean that you have to use these figures to the last decimal place. For many calculations, you can drop a figure or two. Thus, for converting miles to kilometers, you can change 1.609 to 1.61, or even to 1.6, depending on how accurate an answer is required.

In general, the conversions are listed with numbers greater than one. Most people find it easier to work with a number like 39.37 inches (equals one meter), rather than 0.0254 meter (equals one inch). If you want to convert from inches to meters, you may look first for "inches" in the table. Under "inch" it will refer you to "meter." Looking alphabetically under "meter" you will find that 1 meter = 39.37 inches.

Since only one conversion factor is listed for each combination of two units, it is sometimes necessary to multiply by this factor, and at other times to divide by it. Knowing for certain which of these to do troubles some people. There is an easy way to tell which method to use. It is explained in the next chapter.

ENGLISH-METRIC CONVERSIONS

1 acre = 4,046.87 square meters (m²)

acres (see square mile, square kilometer, hectare)

1 atmosphere = 14.696 pounds/square inch (lb/sq in)

1 Btu = 777.98 foot-pounds (ft-lb)

1 Btu = 252.0 gram-calories (g-cal)

Btu (see kilogram-calorie)

bushels (dry) (see cubic meter)

centimeters (see inch, foot, yard)

1 centimeter/second (cm/sec) = 1.9685 feet/minute (ft/min)

centimeters/second (see foot/second)

cubic centimeters (see cubic inch, pint, quart, ounce-fluid)

1 cubic foot (cu ft) = 7.48052 gallons (U.S.) (gal)

1 cubic foot (cu ft) = 28.32 liters (l)

cubic feet (see cubic meter)

1 cubic inch (cu in) = 16.387 cubic centimeters (cm³)

1 cubic meter (m³) = 28.38 bushels (dry) (bu)

1 cubic meter (m³) = 1.308 cubic yards (cu yd)

1 cubic meter (m³) = 35.31 cubic feet (cu ft)

1 cubic yard (cu yd) = 764.54 liters (l)

cubic yards (see cubic meter)

1 dram (avoirdupois) = 1.7718 grams (g)

feet (see meter, furlong)

feet/minute (see meter/minute, centimeter/second)

1 foot (ft) = 30.48 centimeters (cm)

foot-pounds (see Btu)

1 foot/second (ft/sec) = 30.48 centimeters/second (cm/sec)

1 foot-pound (ft-lb) = 1.3558 joules

1 furlong = 40.0 rods

1 furlong = 660.0 feet (ft)

1 gallon (US) (gal) = 3.785 liters (l)

gallons (US) (see cubic foot)

1 gallon of water = 8.3370 pounds of water (at 60°F)

grains (see gram)

1 gram (g) = 15.43 grains

grams (see dram, ounce, pound, ounce-troy)

gram-calories (see Btu)

1 gram/square centimeter (g/cm²) = 2.373 pounds/square foot
(lb/sq ft)

1 hectare = 2.471 acres

hectares (see square mile)

1 horsepower
(550 ft-lb/sec) = 1.014 horsepower (metric)

1 horsepower = 745.7 watts

1 inch (in) = 2.540 centimeters (cm)

1 inch (in) = 25.40 millimeters (mm)

inches (see meter)

joules (see foot-pound)

1 kilogram (kg) = 2.205 pounds (lb)

kilograms (see ton-short, ton-long)

1 kilogram-calorie = 3.968 Btu

kilograms/cubic meter (see pound/cubic foot)

1 kilometer (km) = 0.6214 mile (mi) (ordinary, or statute mile)

1 kilometer (km) = 1,093.6 yards (yd)

kilometers (see mile–nautical, mile–statute)

1 liter (l) = 1.057 quarts (qt)

1 liter (l) = 2.114 pints (pt)

liters (see cubic foot, cubic yard, gallon)

1 meter (m) = 3.281 feet (ft)

1 meter (m) = 39.37 inches (in)

1 meter (m) = 1.094 yards (yd)

1 meter/minute (m/min) = 3.281 feet/minute (ft/min)

1 mile (nautical) = 1.852 kilometers (km)

1 mile (common, or statute) = 1.609 kilometers (km)

milliliters (same as cubic centimeters)

millimeters (see inch)

1 ounce (oz) = 28.3495 grams (g) (this is the common, or
        avoirdupois ounce, of which there are 16 to one pound)

ounce (avoirdupois) (see ounce–troy)

1 ounce (fluid) = 29.57 cubic centimeters (cm³)

                = 29.57 milliliters (ml)

1 ounce (troy) = 31.1035 grams (g)

1 ounce (troy) = 1.09714 ounces (avoirdupois)

1 pint (pt) = 473.2 cubic centimeters (cm³)

pints (see liter)

1 pound = 453.5924 grams (g)

pounds (see kilogram)

1 pound/cubic foot (lb/cu ft) = 16.02 kilograms/cubic meter
        (kg/m³) = 16.02 grams/liter (g/l)

pounds/square foot (see gram/square centimeter)

pounds/square inch (see atmosphere)

pounds of water (see gallon of water)

1 quart (qt) = 946.4 cubic centimeters (cm³)

quarts (see liter)

square centimeters (see square inch, square foot)

square feet (see square meter)

1 square foot (sq ft) = 929.0 square centimeters (cm²)

1 square inch (sq in) = 6.452 square centimeters (cm²)

1 square kilometer (km²) = 247.1 acres

square kilometers (see square mile)

1 square meter (m²) = 10.764 square feet (sq ft)

1 square meter (m²) = 1.196 square yards (sq yd)

square meters (see acre)

1 square mile = 640.0 acres

1 square mile = 2.590 square kilometers (km²)

1 square mile = 259.00 hectares

square yards (see square meter)

rods (see furlong)

1 ton (long) = 2,240 pounds (lb)

1 ton (long) = 1,016 kilograms (kg)

1 ton (long) = 1.016 metric tons

1 ton (metric) = 1.1023 tons (short)

1 ton (short) = 2,000 pounds (lb)

1 ton (short) = 907.185 kilograms (kg)

watts (see horsepower)

1 yard (yd) = 91.44 centimeters (cm)

yards (see meter, kilometer)

13

An Easy Way To Solve Conversion Problems

Converting one English unit to another English unit, or one metric unit to an English unit, can be difficult. There is, however, a method that can be used to simplify these calculations. It is commonly called the "ratio method." I have used this method for many years, and can assure you that it is well worth the small effort required to learn it.

The method is based on the fact that any conversion factor, such as 12 inches = 1 foot, can be expressed as a ratio. It can be written $\left(\dfrac{12 \text{ in}}{1 \text{ ft}}\right)$ or $\left(\dfrac{1 \text{ ft}}{12 \text{ in}}\right)$.

To work problems with this method, it is necessary to change the needed conversion factors into ratios such as these.

Here are the steps that must be followed. They must be done in this order, and *no* step may be omitted.

1. Write down the number that is to be converted and after the number write an abbreviation of the units of the number (ft, in, cm³, etc.).

2. Multiply this number by the proper conversion ratio, so the units of the original number will cancel.

3. Actually draw a line through the units of the original number and one of the units of the ratio to cancel them.

4. Perform the indicated multiplication or division of the two numbers to find the answer to the problem.

5. The units remaining on the left side of the equation are the units of the answer. *Write them down* after the answer.

To illustrate these steps, let's calculate the number of inches in 15 feet. This is a simple problem where you multiply 15 by 12 to get the answer. But let's solve it with the ratio method, following the five steps outlined above.

Step 1. Write: (15 ft).

Step 2. The (15 ft) must be *multiplied* by a ratio, so the units of "ft" will cancel, leaving the units of "in." Here are the two possibilities:

$$(15 \text{ ft}) \left(\frac{1 \text{ ft}}{12 \text{ in}} \right) \quad \text{or} \quad (15 \text{ ft}) \left(\frac{12 \text{ in}}{1 \text{ ft}} \right)$$

The first ratio is *not* correct, since the units of "ft" will not cancel. For units to cancel, one must be in the numerator and the other in the denominator. So the second equation is the correct one.

Step 3. $(15 \text{ ft}) \left(\dfrac{12 \text{ in}}{1 \text{ ft}} \right)$

Steps 4 and 5. $(15 \text{ ft}) \left(\dfrac{12 \text{ in}}{1 \text{ ft}} \right) = 180 \text{ in}$

You can say that this is a lot of extra work when you know that all you need to do is to multiply 15×12 to get the answer. But the method does three things: (1) it indicates that 15 is to be multiplied by 12 to get the answer; (2) it *proves* that this is the correct method, without any thought on your part as to whether you should multiply or divide; (3) it automatically gives the units of the answer so you can look at them and say, "Yes, I wanted the answer to be in inches and it is."

Once the method is understood, it is not necessary to write both ratios and then decide which one to use. You merely look at the original number and its units, and immediately write the correct ratio, so the units of the original number will cancel. To illustrate this, suppose the problem is to calculate the number of seconds in 3.5 hours. Steps 1-5 can be put down all at once, as long as you do them in the proper order. For example:

$$(3.5 \text{ hr}) \left(\frac{60 \text{ min}}{1 \text{ hr}} \right) \left(\frac{60 \text{ sec}}{1 \text{ min}} \right) = 12,600 \text{ sec}$$

In this problem, two ratios were used. The first ratio cancelled "hr" and left "min," and the second ratio cancelled "min" and left "sec." And "sec" is what was asked for. Also, the setup indicates that 3.5 must be multiplied by 60 and again by 60 to give the answer.

Sometimes the numbers in the problem appear in the denominator of the ratio. Then the original number must be divided by the number in the ratio. To illustrate this, let's calculate the number of yards in 3,000 inches. The problem is set up as follows:

$$(3{,}000 \text{ in}) \left(\frac{1 \text{ ft}}{12 \text{ in}}\right) \left(\frac{1 \text{ yd}}{3 \text{ ft}}\right) = 83\tfrac{1}{3} \text{ yd}$$

Here the first ratio cancelled "in" and left "ft." The second ratio cancelled "ft" and left "yd." So the units of the answer are "yd," and that is what was to be calculated. In this case, to get the number of the answer, it is necessary to divide 3,000 by 12 and then divide that answer by 3. (You could also multiply 12 by 3 to give 36 and then divide 3,000 by 36.) Of course, it would be simpler to use the one ratio $\left(\dfrac{1 \text{ yd}}{36 \text{ in}}\right)$ to obtain the answer.

Remember that the original number is *always* multiplied by the conversion ratio, or by two or more ratios. But the numbers of the problem sometimes must be multiplied and sometimes divided by one another. After you set up the ratios and cancel your units, you will see what to do with the numbers.

The ratio method can be used for any problem in which one number must be converted into another. This can be from one number to another in the English system, a metric unit to English, or the reverse.

The beauty of this method is that it allows you to work complex problems that involve a number of conversion factors. It helps when a conversion table doesn't list the conversion that you need. If it is desired to convert original units that can be called "a," into units "d," then it may be possible to find a conversion factor that will convert "a" into "b," a second one that will convert "b" into "c," and finally a third one that will convert "c" into the desired "d."

As an example of the use of multiple ratios, a problem might read,

"How many kilometers are there in 10 miles?"

We could do this easily with one ratio that converts miles to kilometers. But suppose we know only the following:

1 mile = 5,280 feet
12 inches = 1 foot
2.54 centimeters = 1 inch
100 centimeters = 1 meter
1,000 meters = 1 kilometer

Now, if we do it right, we can convert miles to feet, feet to inches, inches to centimeters, centimeters to meters, and finally meters to kilometers. Here is how it is done:

$$(10 \text{ mi})\left(\frac{5,280 \text{ ft}}{1 \text{ mi}}\right)\left(\frac{12 \text{ in}}{1 \text{ ft}}\right)\left(\frac{2.54 \text{ cm}}{1 \text{ in}}\right)\left(\frac{1 \text{ m}}{100 \text{ cm}}\right)\left(\frac{1 \text{ km}}{1,000 \text{ m}}\right)$$

So the answer is:

$$\frac{(10)(5{,}280)(12)(2.54)}{(100)(1{,}000)} = 16.09 \text{ km}$$

Very few problems involve this many ratios, but this shows what can be accomplished.

Many things are conversion factors that you may not have thought of as such. Anything with a "per" in the expression is a conversion factor. For example, eggs may be 60 cents *per* dozen. "Per" always means *one*. Written as a ratio, this reads:

$$\frac{60\cancel{}}{1 \text{ doz}} \quad \text{or} \quad \frac{1 \text{ doz}}{60\cancel{}}$$

If eggs are 60 cents per dozen, what is the cost of 25 eggs? The problem is solved as follows:

$$(25 \text{ eggs}) \left(\frac{1 \text{ doz}}{12 \text{ eggs}} \right) \left(\frac{60\cancel{}}{1 \text{ doz}} \right) = 125\cancel{} \text{ or } \$1.25$$

The density of material is expressed in the metric system in grams per cubic centimeter (g/cm^3). This is a conversion factor between the weight in grams and the volume in cubic centimeters that the weight occupies. What, for example, is the weight of 150 cubic centimeters of a material with a density of 3.2 g/cm^3? It is:

$$(150 \text{ cm}^3) \left(\frac{3.2 \text{ g}}{1 \text{ cm}^3} \right) = 480 \text{ g}$$

Paper for printing is sold on "basis weight." In the metric system, basis weight is the weight in grams of one sheet of the paper that measures one meter by one meter, the area of which equals one square meter. So the units of basis weight are grams per square meter (g/m²). This, too, is a conversion factor. If a particular paper has a basis weight of 90 g/m², the ratio is:

$$\left(\frac{90 \text{ g}}{1 \text{ m}^2}\right) \quad \text{or} \quad \left(\frac{1 \text{ m}^2}{90 \text{ g}}\right)$$

Let's calculate the weight in kilograms (kg) of 10,000 sheets of paper 40 × 60 centimeters (cm). First we must get the area of one sheet in square meters: 40 cm × 60 cm becomes 0.40 m × 0.60 m = 0.24 m². This is a conversion factor since it equals 0.24 m² *per* sheet. The ratio is either:

$$\left(\frac{0.24 \text{ m}^2}{1 \text{ sheet}}\right) \quad \text{or} \quad \left(\frac{1 \text{ sheet}}{0.24 \text{ m}^2}\right)$$

Now we have the information necessary to solve the problem. It is:

$$(10{,}000 \text{ sheets})\left(\frac{0.24 \text{ m}^2}{1 \text{ sheet}}\right)\left(\frac{90 \text{ g}}{1 \text{ m}^2}\right)\left(\frac{1 \text{ kg}}{1{,}000 \text{ g}}\right) = 216 \text{ kg}$$

In all of the problems solved thus far, the original unit, whether it was feet, inches, eggs, or sheets, was *not* a conversion factor in itself. One of the rules of the ratio method is that if you start with such a number, it can be multiplied by any required number of conversion factors, but the answer will still be a number that is not itself a conversion factor. For example, we started with 10,000 sheets of paper and we ended with 216 kilograms, neither of which is a conversion factor.

But if the initial unit *is* a conversion factor, then it may be multiplied by any other necessary conversion factors (in the form of ratios) and the answer will always be a conversion factor.

Consider this example: On a trip to Norway, I found that grapes were selling for 6 kroner per kilogram. The kroner is a Norwegian coin worth close to 15 U.S. cents. This is also a conversion factor—15 cents per kroner. How much did the grapes cost in cents per pound?

When a problem begins with a conversion factor, there are two units to convert. In this case, kroners must be converted to cents. Also, kilograms must be converted to pounds. In working the problem, it doesn't matter which one is converted first. The answer is exactly the same:

$$\left(\frac{6 \text{ Kr}}{1 \text{ kg}}\right)\left(\frac{15 \text{ cents}}{1 \text{ Kr}}\right)\left(\frac{1 \text{ kg}}{2.2 \text{ lb}}\right) = \frac{41 \text{ cents}}{\text{lb}}$$

All units cancel, except "cents" in the numerator and "lb" in the denominator. So the units of the answer are *cents per pound*. Any unit in the denominator is read as "per," so the answer is about 41 cents *per* pound.

If you stop to think about it, many figures we deal with are conversion factors. For example: $5.00/hr (for labor); 24 hours/day; 52 weeks/year; 45 cents/gallon (of gasoline); $8.00/sq yd (of carpeting). These and all other conversion factors can be changed into ratios for problem solving.

PROBLEMS

All of the problems given below are to be solved by the ratio method. No conversion factors are given. Some are common knowledge, some you have just learned, and others may be found in the English-Metric Conversion Table on pages 78–81.

1. If a certain grade of carpet has been selling for $9 per square yard, what should the price be per square meter?

2. If a quart of milk costs 30 cents, what should a liter of milk cost?

3. If two cities are 80 kilometers apart and you can drive 90 kilometers per hour driving between them, how many minutes will it take to drive the distance between them?

4. If gasoline has been 55 cents per gallon, what should the cost be for 30 liters?

5. Calculate the number of seconds in a year. Use 365 days for one year.

6. If a printing press is moving paper through at a speed of 1,000 feet per minute, what is the paper speed in centimeters per second?

7. A certain kind of steel has a density of 7.7 grams per cubic centimeter. What is its density in pounds per cubic foot? (Hint—start this problem with the ratio: $\dfrac{7.7 \text{ g}}{1 \text{ cm}^3}$. Then use other ratios to convert grams to pounds and cubic centimeters to cubic feet. There is no figure in the table for converting cubic centimeters to cubic feet. So you must figure how to do it with two or more other ratios.)

Answers to problems begin on page 111.

14

Using Powers of Ten

In a problem in which several numbers must be multiplied and divided, it is sometimes difficult to put the decimal point in the right place in the answer. Using powers of 10 can make this easier. When you use powers of 10, the decimal point in very large or very small numbers is moved in the right (or left) direction to make a more workable number—one between 1.0 and 9.9 that is multiplied by 10 to a given power.

For example, a number like 352 can be changed to 3.52×10^2. The power of 10, which is 2 in this case, is the same as the number of places the decimal point has been moved. Similarly, 0.000175 can be changed to 1.75×10^{-4}. Since 1.75 is 10,000 times bigger than 0.000175, the power of 10 must be -4.

The rules for using powers of 10 are:

1. When the decimal point is moved to the left, the power of 10 is a positive number.

2. When the decimal point is moved to the right, the power of 10 is a negative number.

3. The power number indicates the number of places the decimal point is moved.

Here are a few more examples to make this clear:

$$78,000 = 7.8 \times 10^4$$
$$0.0275 = 2.75 \times 10^{-2}$$
$$10.4 = 1.04 \times 10^1$$

This list shows what some of the powers of 10 mean:

10^6	is	1,000,000.
10^5	is	100,000.
10^4	is	10,000.
10^3	is	1,000.
10^2	is	100.
10^1	is	10.
10^0	is	1.
10^{-1}	is	0.1
10^{-2}	is	0.01
10^{-3}	is	0.001
10^{-4}	is	0.0001 (one ten-thousandth)
10^{-5}	is	0.00001 (one hundred-thousandth)
10^{-6}	is	0.000001 (one millionth)

The idea of this method is to convert both large and small numbers to numbers between 1.0 and 9.9 that can be easily multiplied or divided. Then you operate *separately* on the powers of 10 to find where the decimal place goes in the answer.

A typical problem might be:

$$(34,000)(0.025)$$

Using powers of 10, the problem would be solved as follows:

$$(3.4 \times 10^4)(2.5 \times 10^{-2}) = 8.50 \times 10^2 = 850$$

When 3.4 and 2.5 are multiplied, they equal 8.50. Then 10^4 is multiplied by 10^{-2}, to give 10^2. The number 10^2 means 10 squared, and 10 squared is 100. So the final number is 100 times as big as 8.50 or 850. The decimal has been moved two places to the right. Actually 8.50×10^2 is also a correct answer, but to eliminate the 10^2 the 8.50 must be changed to 850.

To multiply 10 to some power by 10 to some other power, the power numbers must be added *algebraically*.

For those who are not familiar with algebraic addition, here are the rules:

Rule 1. If two positive numbers are added, the answer is a positive number:

$$(2 + 3 = 5)$$

Rule 2. If two negative numbers are added, the answer is a negative number:

$$(-2 + -3 = -5)$$

Rule 3. If one number is positive and the other is negative, subtract the smaller number from the larger number. When the larger number is positive, the answer is positive. When the larger number is negative, the answer is negative:

$$(-2 + 3 = 1)$$
$$(-4 + 2 = -2)$$

Here are some examples and the rules that apply to them:

$$10^2 \times 10^3 = 10^5 \quad \text{(Rule 1)}$$
$$10^{-2} \times 10^{-3} = 10^{-5} \quad \text{(Rule 2)}$$
$$10^{-2} \times 10^3 = 10^1 \quad \text{(Rule 3)}$$
$$10^2 \times 10^{-4} = 10^{-2} \quad \text{(Rule 3)}$$

In some problems, division is required. In this case, the power number in the denominator must be *subtracted* algebraically from the power number in the numerator. To do this, change the sign of the power in the denominator and add it algebraically to the power in the numerator.

Consider these examples:

$$\frac{10^4}{10^3} = 10^1$$

$$\frac{10^{-3}}{10^{-1}} = 10^{-2}$$

$$\frac{10^3}{10^{-2}} = 10^5$$

$$\frac{10^{-3}}{10^2} = 10^{-5}$$

In the first division example, this becomes $4 - 3$, which is 1, so the answer is 10^1. In the second example, it becomes $-3 + 1$, which is -2. In the third example, it is $3 + 2$, or 5. In the fourth example, it is $-3 + -2$, or -5.

Let's see how a combination multiplication and division problem can be worked by this method. Consider the problem:

$$\frac{(0.028)(4,500)}{(0.18)}$$

These figures are manipulated as follows:

$$\frac{(2.8 \times 10^{-2})(4.5 \times 10^3)}{(1.8 \times 10^{-1})}$$

The next step is to multiply 2.8 by 4.5 and then to divide that answer by 1.8. This gives 7.

Now we must operate on the powers of 10 by adding -2 and $+3$, and then subtracting -1. It reads as follows:

$$-2 + 3 - (-1) = -2 + 3 + 1 = 2$$

This means that 10 will have a power of 2 in the answer. The answer is 7.0×10^2, or 700.

Sometimes when the numbers between 1.0 and 9.9 are multiplied or divided, the answer is more than 10, or less than one. Here are two examples:

$$(2.1 \times 10^3)(8.0 \times 10^{-1}) = 16.8 \times 10^2 = 1,680$$

$$\frac{2.5 \times 10^3}{5.0 \times 10^2} = 0.5 \times 10^1 = 5.0$$

More complex problems are not much more difficult than the ones that have been given. Such as:

$$\frac{(2.0 \times 10^2)\,(8.0 \times 10^{-3})}{(1.0 \times 10^3)\,(4.0 \times 10^{-1})} = \frac{16.0 \times 10^{-1}}{4.0 \times 10^2}$$
$$= 4.0 \times 10^{-3} = 0.004$$

Here are the two rules to help you eliminate a given power of 10 in the answer:

1. If the power of 10 is a positive number, move the decimal point to the right the number of places in the power.

2. If the power of 10 is a negative number, move the decimal point to the left the number of places in the power.

Here are some examples:

$$8.2 \times 10^4 = 82{,}000$$
$$3.7 \times 10^1 = 37$$
$$7.8 \times 10^{-2} = 0.078$$
$$15.6 \times 10^{-3} = 0.0156$$
$$4.5 \times 10^0 = 4.5$$

In the last example, 10^0 is actually 1.0, so the number 4.5 remains the same.

If you are not familiar with algebraic addition and subtraction, this will probably be a difficult method for you to use. It is a particularly handy method, however, if you are performing multiplication and division with a slide rule. A slide rule tells you nothing about the position of the decimal point. You set 4.3, 4,300 and 0.0043 on exactly the same place on a slide rule. So some method must be used to establish the decimal point, and this method will serve that purpose.

PROBLEMS

Convert the figures in the following problems into small numbers times 10 to the proper power. Then carry out the multiplication and division to obtain the answer to one decimal place in the small number part of the answer.

1. $(12,000)(3.5)(0.0025)$

2. $\dfrac{15,000}{0.75}$

3. $\dfrac{0.0075}{1,500}$

4. $\dfrac{(23,000)(0.056)}{150}$

5. $\dfrac{(450,000)(5.5)}{2,500}$

6. $\dfrac{(0.170)(0.0035)}{0.50}$

7. If you inherited $1,200,000 (a dream, perhaps!) and had it invested with an average return of 6.2% per year, what would your annual income be? (Hint: For use in a problem, 6.2% becomes 0.062.)

Answers to the problems begin on page 112.

ANSWERS TO PROBLEMS

1. a. 0.550 kg
   b. 55,000 mg
   c. 0.247 g
   d. 5,480 g
   e. 0.548 metric ton
   f. 25,800 kg

2. a. 2.42 kg   2,420 g
       0.37 kg   (or)   370 g
       ─────────   ─────────
       2.79 kg = 2,790 g   2,790 g = 2.79 kg

   b.   350 kg
       5150 kg
       ─────────
       5500 kg = 5.500 metric ton

     (or)   0.350 metric ton
            5.15   ″   ″   ″
            ─────────
            5.500 metric ton = 5500 kg

Quick Reference Chart for English-Metric Conversions

UNITS OF WEIGHT

1 ounce = 28 grams
1 pound = 500 grams (a little more
  than one pound; 450 grams comes
  closer to a pound)
1 kilogram = 2.2 pounds
A metric ton = about 10% more than
  an English short ton (the ton most
  people use, of 2,000 lb.).

[NOTE: The purpose of this table is to help you to learn how big some of the metric units are. When two units below are said to be equal, it must be remembered that they are only approximately equal. For best results, you should stop looking at this table just as soon as possible.]

UNITS OF LENGTH

$2\frac{1}{2}$ centimeters = 1 inch
30 centimeters = 1 foot
1 meter = 1 yard plus about 3 inches
1 kilometer = 0.6 mile

AREAS

1 square inch = $6\frac{1}{2}$ times 1 square centimeter
1 square yard = a little less than 1 square meter
1 hectare = $2\frac{1}{2}$ acres
1 square mile = 2.6 square kilometers

VOLUME

1 fluid ounce = 30 cubic centimeters (or 30 milliliters)
1 liter = a little more than 1 quart (about 5% more)
4 liters = a little more than 1 gallon
1 cubic meter = a little more than 1 cubic yard

COMPARISON OF TWO TEMPERATURE SCALES

Degrees Celsius	Degrees Fahrenheit	
100	212	(water boils here)
37	98.6	(body temperature)
30	86	
20	68	
10	50	
0	32	(water freezes here)

c. 2.84 g   2,840 mg
   0.25 g   (or)   250 mg
   3.09 g = 3,090 mg   3,090 mg = 3.09 g

3. 500 g. is more than one pound. 1000 g. (one kg) is about 2.2 lb, so 500 g. is about 1.1 lb.

4. There are about 2.2 lb in one kilogram. So

$$176 \text{ lb} = \frac{176}{2.2} = 80 \text{ kg}$$

5. $\dfrac{160 \text{ lb}}{2.2 \text{ kg/lb}} = 72.7 \text{ kg}$

   (72.7 kg)(25 calories/kg) = 1817.5 calories

6. The smaller package is the better buy. Three small packages cost 3 × 15, or 45¢, and contain 3 × 170, or 510 g. This 510 g. is greater than the 425 g. weight of the bigger package, which costs 47¢ — or 2¢ more than 45¢.

   For an accurate comparison, the number of grams that you can get for one cent can be calculated. With the smaller package this is $\dfrac{170 \text{ g}}{15 \text{ ¢}} = 11.3$ g per cent. For the larger package it is $\dfrac{425 \text{ g}}{47 \text{ ¢}} = 9.05$ g per cent.

   So a cent buys more grams of raisins with the smaller package.

7. It should be $(15¢/\text{lb})(2.2 \text{ lb/kg}) = 33¢/\text{kg}$.

I hope you did fairly well in getting the correct answers to these problems. But most people will make some mistakes the first time they try them. Don't stop! First, study the correct answers, compare them with what you did, and see where you made your mistakes. If you still don't understand the correct method, then go back and study the chapter on metric units of weight again, until you do understand. Finally, wait a week or so, and then try working the problems again. I hope you make "100" the second time you try! When you can work these—or similar—problems correctly and rapidly, with all of the conversion factors in your head, then you will at least have a start on "thinking metric".

This same procedure should be followed when you work the problems that follow these. Keep at each set until you achieve perfection. Good luck!

Chapter 2

1. a. 35.7 cm (there are 10 millimeters in one centimeter)
   b. 0.357 m
   c. 428 mm
   d. 0.428 m
   e. 498 cm
   f. 0.730 km
   g. 8,130 m

2. $(60 \text{ km})(0.6 \text{ mi/km}) = 36$ miles

3. $3,400 \text{ m} = 3.4 \text{ km}$
   $(3.4 \text{ km})(0.6 \text{ mi/km}) = 2.04$ miles

4. 10% of 400 is 40 $400 + 40 = 440$ yards

5. 2 m 60 cm = 260 cm
   3 m 75 cm = 375 cm
   $\overline{}$
                635 cm, or 6.35 m

   (or)   2.60 m
           3.75 m
           $\overline{}$
           6.35 m, which is 635 cm

6. 4 m 15 cm =   415 cm
   2 m 85 cm = -285 cm
              $\overline{}$
               130 cm = 1.30 m

   (or)   4.15 m
           -2.85 m
           $\overline{}$
            1.30 m = 130 cm

7. 2 m 31 cm = 231 cm $\left(\frac{1}{3}\right)(231) = 77$ cm

   So each length should be 77 cm.

   $\left(\frac{1}{2}\right)(231) = 115.5$ cm. This is the halfway distance.

   It could be measured as one meter 15.5 cm.

8. (a) $\dfrac{89 \text{ cm}}{2\frac{1}{2} \text{ cm/in}} = 35.6$ inches

$$\frac{58 \text{ cm}}{2\frac{1}{2} \text{ cm/in}} = 23.2 \text{ inches}$$

Thus the answer to (a) is: 35.6—23.2—35.6″.

(b) $\dfrac{170 \text{ cm}}{2\frac{1}{2} \text{ cm/in}} = 68$ inches, or 5 feet 8 inches

(c) $(53 \text{ kg})(2.2 \text{ lb/kg}) = 116.6$ lb

9. $(200 \text{ km})(0.6 \text{ mi/km}) = 120$ miles

10. $(110 \text{ km/hr})(0.6 \text{ mi/km}) = 66$ miles per hour

(Re-read the suggestions at the end of the Chapter 1 problems, and follow the same procedure with these.)

Chapter 3

1. a. 17,500 cm² (1 m² = 10,000 cm² so decimal point must be moved four places)
   b. 24 ares
   c. 4.50 hectares
   d. 0.0450 km² (1 square kilometer = 10,000 ares, so decimal point must be moved four places)
   e. 245 hectares
   f. 485 ares
   g. 2,000 m²

2. a. $(0.5 \text{ km})(1.5 \text{ km}) = 0.75 \text{ km}^2$
   b. $(0.75 \text{ km}^2)(100 \text{ hectares/km}^2) = 75$ hectares

b. (second method) $(500 \text{ m})(1500 \text{ m}) = 750,000 \text{ m}^2$

$$\frac{750,000 \text{ m}^2}{10,000 \text{ m}^2/\text{hectare}} = 75 \text{ hectares}$$

(To divide by 10,000 the decimal point in 750,000 must be moved four places to the left.)

c. $(75 \text{ hectares})(100 \text{ ares/hectare}) = 7,500 \text{ ares}$

c. (second method) $(0.75 \text{ km}^2)(10,000 \text{ ares/km}^2) = 7,500 \text{ ares}$

3. $(4.25 \text{ m})(3.45 \text{ m}) = 14.6625 \text{ m}^2 = 146,625 \text{ cm}^2$

4. 10 ft = 120 in
$(120 \text{ in})(2.5 \text{ cm/in}) = 300 \text{ cm} = 3 \text{ m}$
15 ft is $1\frac{1}{2}$ times 10 ft. So 15 ft is $1\frac{1}{2}$ times 3 m, or 4.5 m
Floor area $= (3 \text{ m})(4.5 \text{ m}) = 13.5 \text{ m}^2$
Cost of carpeting $= (13.5 \text{ m}^2)(\$9.00/\text{m}^2) = \$121.50$

5. $(150 \text{ cm})(110 \text{ cm}) = 16,500 \text{ cm}^2$ of wall area
Area of one tile is $(11 \text{ cm})(11 \text{ cm}) = 121 \text{ cm}^2$

Minimum number of tile needed $= \dfrac{16,500}{121} = 137$ tile

6. $(50 \text{ hectares})(2.5 \text{ acres/hectare}) = 125 \text{ acres}$

7. Two walls have an area of $4.25 \text{ m} \times 2.8 \text{ m}$ or 11.9 m^2 each. So area of these two walls is $(11.9)(2) = 23.8 \text{ m}^2$.

The other two walls have an area of 5.60 m $\times$ 2.8 m or 15.68 m² each. So area of these two walls is $(15.68)(2) = 31.36$ m².

Total wall area $= 23.8 + 31.36 = 55.16$ m².
Area of each door is $(1$ m$)(2.30$ m$) = 2.30$ m² or 4.60 m² for the two doors.

Area of window is $(1.50$ m$)(2$ m$) = 3.00$ m²

Total area of doors and window $= 4.60 + 3.00 = 7.60$ m²

$$\begin{array}{l} 55.16 \text{ m}^2 \text{ (total wall area)} \\ \underline{-7.60 \text{ m}^2 \text{ (doors and window)}} \\ 47.56 \text{ m}^2 \text{ of wallpaper needed} \end{array}$$

This amount is minimum and doesn't allow for fitting.

8. $(3.5$ yd$)(3$ ft/yd$) = 10.5$ ft

$$\frac{45 \text{ in}}{12 \text{ in/ft}} = 3.75 \text{ ft}$$

Area of material needed $= (10.5$ ft$)(3.75$ ft$)$
$\qquad\qquad\qquad\qquad\quad = 39.38$ sq ft.

$$\frac{39.38 \text{ sq ft}}{10.76 \text{ sq ft/m}^2} = 3.66 \text{ m}^2$$

Material length needed, in meters $= \dfrac{3.66 \text{ m}^2}{1.10 \text{ m (width)}}$
$\qquad\qquad\qquad\qquad\qquad\qquad\quad = 3.33$ meters

9. $(13.4 \text{ cm})(18.5 \text{ cm}) = 247.90 \text{ cm}^2$

Some of these problems will be difficult for many people. At least the conversions that are involved will be good practice. Don't become discouraged if you couldn't work some of these problems that relate to areas. How often do you calculate areas in the English system? Probably not very often. So, if you have to calculate an area once in a while in the metric system, you can come back and read this chapter again.

Chapter 4

1. $(6 \text{ cm})(6 \text{ cm})(6 \text{ cm}) = 216 \text{ cm}^3$

2. $(44 \text{ cm})(32 \text{ cm})(28 \text{ cm}) = 39,424 \text{ cm}^3 = 39.424 \text{ liters}$

3. (a) $(44 \text{ cm})(32 \text{ cm}) = 1,408 \text{ cm}^2$, or $2,816 \text{ cm}^2$ for two sides, this size

   $(44 \text{ cm})(28 \text{ cm}) = 1,232 \text{ cm}^2$, or $2,464 \text{ cm}^2$ for two sides, this size

   $(32 \text{ cm})(28 \text{ cm}) = 896 \text{ cm}^2$, or $1,792 \text{ cm}^2$ for two sides, this size

   2,816
   2,464
   1,792
   ‾‾‾‾‾
   $7,072 \text{ cm}^2 = $ total surface area

(b) (0.44 m)(0.32 m) = 0.1408 m², or 0.2816 m² for these two sides. These figures are exactly the same as the square centimeter figures, except the decimal point is moved four places to the left. So we get:

0.2816
0.2464
0.1792
0.7072 m² = total surface area

4. 1500 ml

5. $\dfrac{150}{30} = 5$ fl oz

6. 1 m³ = 1000 l   Thus 10,500 l = 10.5 m³

7. A liter is about 5% bigger than a quart. Five per cent is 0.05   (10)(0.05) = 0.5
Thus 10 liters are about 10.5 quarts

8. A little less than 10 × 4, or 40 liters.

Chapter 5

1. (a) 450 g   (b) 0.450 kg

2. Weight of water = 475 − 100 = 375 g
Volume of container is (a) 375 cm³   (b) 0.375 l

3. (a) 2.47 kg   (b) 2,470 g (one liter of water weighs one kilogram)

4. (a) Volume of bar = (5 cm)(3 cm)(160 cm) = 2,400 cm³ (2,400)(2.70) = 6,480 g

   (b) 6.480 kg

5. (a) $\dfrac{12.340 \text{ kg}}{8.92 \text{ kg/l}} = 1.38 \text{ l}$ (to 2 decimals)

   (b) 1,380 cm³

6. (a) Volume of aquarium = (30 cm)(40 cm)(20 cm) = 24,000 cm³
   Weight of water = 24,000 g

   (b) 24.000 kg

7. (a) Volume of water in pool = (7 m)(14 m)(1.4 m) = 137.2 m³
   137.2 m³ = 137,200 l
   Weight of water = 137,200 kg

   (b) 137.200 metric tons

   (c) A metric ton is about 10% bigger than a US short ton. Ten per cent of 137.2 is 13.7 (to one decimal). So the number of US short tons is approximately 137.2 + 13.7 = 150.9

Chapter 6

88°F
1. 84°F
   72°F
   66°F
   54°F

48°F
36°F
30°F

2. 95°F
77°F
59°F
41°F
23°F

3. Yes, she has a little fever, since 37°C is normal body temperature.

Chapter 11

1. Area $= \frac{1}{2}$ bh $= \frac{1}{2}$ (40)(25) $= 500$ cm^2

2. Area $=$ b $\times$ h $=$ (30)(15) $= 450$ cm^2

3. Area $= \pi$ r$^2 =$ (3.14)(8^2) $= 3.14 \times 64 = 200.96$ cm^2

4. Area $= 4\pi$r$^2 =$ (4)(3.14)(8^2) $=$ (4)(3.14)(64) $=$ 803.84 cm^2

5. Volume $= \pi$r^2 h
   (a) (3.14)(7^2)(110 cm) $=$ (3.14)(49)(110) $=$ 16,924.6 cm^3
   (b) There are 1000 liters in one cubic meter, and 1000 cubic centimeters in one liter. So there are 1000 $\times$ 1000, or 1,000,000 cubic centimeters in one cubic meter. This is 10^6. So the answer in (a) can be converted by moving the decimal

point six places to the left, to give 0.0169246 m³. The same answer can be obtained by converting all the initial measurements to meters. Thus we get:

$$(3.14)(.07 \text{ m})(.07 \text{ m})(1.1 \text{ m}) =$$
$$(3.14)(7.0 \times 10^{-2})(7.0 \times 10^{-2})(1.1) =$$
$$169.246 \times 10^{-4} = 0.0169246 \text{ m}^3$$

Chapter 13

1. $\left(\dfrac{\$9.00}{1 \text{ sq yd}}\right)\left(\dfrac{1.196 \text{ sq yd}}{1 \text{ m}^2}\right) = \$10.76/\text{m}^2$

2. $\left(\dfrac{30 \text{ cents}}{1 \text{ qt}}\right)\left(\dfrac{1.057 \text{ qt}}{1 \text{ liter}}\right) = 31.7$ cents/liter

3. $(80 \text{ km})\left(\dfrac{1 \text{ hr}}{90 \text{ km}}\right)\left(\dfrac{60 \text{ min}}{1 \text{ hr}}\right) = 53.3$ min

4. $(30 \text{ liters})\left(\dfrac{1 \text{ gal}}{3.785 \text{ liters}}\right)\left(\dfrac{55 \text{ cents}}{1 \text{ gal}}\right)$
   $= 436 \text{ cents} = \$4.36$

5. $(365 \text{ days})\left(\dfrac{24 \text{ hr}}{1 \text{ day}}\right)\left(\dfrac{60 \text{ min}}{1 \text{ hr}}\right)\left(\dfrac{60 \text{ sec}}{1 \text{ min}}\right) =$

   $(3.65 \times 10^2)(2.4 \times 10^1)(6.0 \times 10^1)(6.0 \times 10^1) =$
   $315.36 \times 10^5 = 31,536,000$ sec

6. $\left(\dfrac{1000 \text{ ft}}{1 \text{ min}}\right)\left(\dfrac{1 \text{ min}}{60 \text{ sec}}\right)\left(\dfrac{30.48 \text{ cm}}{1 \text{ ft}}\right) = 508$ cm/sec

Or, from the conversion tables, we find:
1 cm/sec = 1.97 ft/min. So we can write:

$$(1000 \text{ ft/min})\left(\frac{1 \text{ cm/sec}}{1.97 \text{ ft/min}}\right) = 508 \text{ cm/sec}$$

7. $\left(\dfrac{7.7 \text{ g}}{1 \text{ cm}^3}\right)\left(\dfrac{1000 \text{ cm}^3}{1 \text{ liter}}\right)\left(\dfrac{28.3 \text{ liters}}{1 \text{ cu ft}}\right)\left(\dfrac{1 \text{ lb}}{453.6 \text{ g}}\right) =$
$\dfrac{(7.7)(1000)(28.3)}{453.6} = 480.4 \text{ lb/cu ft}$

Chapter 14

1. $(12,000)(3.5)(0.0025) =$
$(1.2 \times 10^4)(3.5)(2.5 \times 10^{-3}) = 10.5 \times 10^1 = 105$

2. $\dfrac{15,000}{0.75} = \dfrac{1.5 \times 10^4}{7.5 \times 10^{-1}} = 0.2 \times 10^5 = 20,000$

3. $\dfrac{0.0075}{1500} = \dfrac{7.5 \times 10^{-3}}{1.5 \times 10^3} = 5.0 \times 10^{-6}$
$= 0.000005$

4. $\dfrac{(23,000)(0.056)}{150} = \dfrac{(2.3 \times 10^4)(5.6 \times 10^{-2})}{1.50 \times 10^2} =$
$8.6 \times 10^0 = 8.6$

5. $\dfrac{(450,000)(5.5)}{2,500} = \dfrac{(4.5 \times 10^5)(5.5)}{2.5 \times 10^3} =$
$9.9 \times 10^2 = 990$

6. $\dfrac{(0.170)(0.0035)}{0.50} = \dfrac{(1.7 \times 10^{-1})(3.5 \times 10^{-3})}{5.0 \times 10^{-1}} =$

   $1.2 \times 10^{-3} = 0.0012$

7. Income $= (\$1,200,000)(0.062) =$

   $(\$1.2 \times 10^6)(6.2 \times 10^{-2}) = \$7.44 \times 10^4 = \$74,400$

   per year

INDEX

a, meaning of, 5
Acres
  relation to hectares, 27, 31
Areas, metric, 24–33
  answers to problems
      involving, 104–107
  in square centimeters, 25
  in square meters, 26
  of triangle, circle, etc, 72–76
  problems relating to, 31–33
  problems relating to, of
      triangle, circle and
      parallelogram, 76;
      answers, 110
Ares
  relation to hectare, 27
  relation to square meters, 27

Body temperature, 52
Box, rectangular
  calculation of volume of, 74

c, meaning of, 62
Celsius thermometer
  body temperature on, 52
  equation for conversion of, to
      Fahrenheit, 70–71

freezing and boiling point of
    water on, 50
  problems in converting
      readings of, to Fahrenheit,
      54; answers, 109–110
  relation of, to Fahrenheit
      thermometer, 50–54,
      69–71
centi-
  meaning of, 5, 9, 62
Centimeters
  how to convert to meters,
      17–18
  how to read tenths of, 14, 16
  relation to
      foot, 14, 21
      inch, 14, 21
      meter, 14, 21
      millimeters, 21
  sides of liter cube in, 35
  use of, to calculate areas,
      25
Circle
  calculation of area of, 73
cm, meaning of, 5
cm², meaning of, 5, 25
cm³, meaning of, 5, 35

Computers
    processing speeds of,
        expressed in
        nanoseconds, 63
Cone
    calculation of volume of, 75
Conversion table
    English to metric, 77–81
    metric to metric, 64–68
Cube
    calculation of volume of, 74
Cubic centimeters
    definition of, 41
    relation to
        fluid ounces, 39, 41
        grams of water, 44
        liters, 35, 37, 40
        milliliters, 38, 41
    volume of cup in, 39
    drinking glass in, 39
Cubic meters
    definition of, 39, 41
    calculation of volume in,
        39–40
    relation to
        cubic yards, 41
        liters, 39, 41
Cubic yards
    relation to cubic meters, 41
Cup
    average volume of, in cubic
        centimeters, 39
Cylinder
    calculation of volume of, 75

d, meaning of, 62

da, meaning of, 62
deci-
    meaning of, 62
deciliter, 64
Decimal point
    general rules for moving of, 8
    how to move, to convert
        centimeters to meters,
            17–18
        hectares to ares, 30
        hectares to square
            kilometers, 31
        hectares to square meters,
            27, 30
        kilometers to meters, 19
        one metric weight to
            another, 7–8
        square meters to ares, 30
    movement of, related to
        power of 10, 64–65
decimeter, 62
deka-
    meaning of, 62
dekameter, 62
Density
    definition of, 44
    of some common materials,
        45
    two ways of expressing, 46
    use of, in calculating volume
        from weight, 47–48
    use of, in calculating weight
        from volume, 45–46
Drinking glass
    average volume of, in cubic
        centimeters, 39

English-metric conversion
        table, 77-81

Fahrenheit thermometer
    body temperature on, 52
    equation for conversion of,
        to Celsius, 71
    freezing and boiling point of
        water on, 50
    relation of, to Celsius
        thermometer, 50-54,
        69-71
Fluid ounces (See Ounces, fluid)
Foot
    relation to centimeters, 14, 21
Fractions
    not used in metric system, 6

g, meaning of, 5
G, meaning of, 61
Gallons
    relation to liters, 37, 41
giga-
    meaning of, 61
Grams
    how to convert to kilograms, 7
    reading of scales of, 6
    relation to
        kilograms, 4
        ounces, 4, 6, 10
        pound, 10
    weight of coins in, 4

h, meaning of, 62
ha, meaning of, 5
Hectares
    relation to
        acres, 27, 31
        ares, 27
        square kilometers, 28
        square meters, 27
hecto-
    meaning of, 62
hectogram, 64
hectoliter, 64
Hexagon, regular
    calculation of area of, 74

Inches
    relation to centimeters, 14, 21
    meter, 13

k, meaning of, 62
kg, meaning of, 5
kilo-
    meaning of, 5, 9, 62
Kilograms
    how to convert to grams, 7
    metric tons, 7
    reading on scales of, 6
    relation to
        grams, 4, 6
        metric ton, 4, 6
    pounds, 4, 6, 10
Kilometer
    relation to meters, 14, 19, 21
    mile, 19
Kilometers per hour
    relation of, to miles per hour,
        19
kilowatt, 62
km, meaning of, 5

km², meaning of, 5

ℓ, meaning of, 5
Lengths, metric, 13–23
    adding of, 17–19
    answers to problems
        involving, 102–104
    dividing of, 20
    key conversions of, to
        English lengths, 21
    problems involving, 22–23
Lenses
    focal length of, in metric
        units, 17
Light
    wave lengths of, expressed in
        nanometers, 63
Liters
    relation to
        cubic centimeters, 35, 40
        cubic meters, 39, 41
        gallons, 37, 41
        kilograms of water, 43
        milliliters, 38, 40
        quarts, 37, 41

M, meaning of, 62
m, meaning of, 5, 62
m², meaning of, 5
m³, meaning of, 5
mega-
    meaning of, 62
megacycle, 62
Meters
    how to convert to centimeters,
        17–18

measurement of length and
        width of room in, 16
relation to
    are, 27
    centimeters, 14, 21
    cubic meters, 39
    hectare, 27
    inches, 13
    kilometer, 14, 21
    millimeters, 14, 21
    millimeters, 14, 21
    square kilometers, 28
    yard, 13, 21
    use of, to calculate areas,
        27
Meter stick
    how divided, 16
    relation of, to yard stick, 13
    use of, to measure lengths, 16
Metric Areas (see Areas, Metric)
Metric lengths (see Lengths,
        metric)
Metric-metric conversion table,
        64–68
Metric prefixes (see Prefixes,
        metric)
Metric system
    conversion table of, 64
    learning to think in, 55–57
    used in photography, 17
Metric ton (see Ton, metric)
Metric Volume (See Volume,
        metric)
Metric Weights (see Weights,
        Metric)
mg, meaning of, 5

Micro-
    meaning of, 62
micrometer, 62
microsecond, 63
Mile
  relation to kilometer, 19
Miles per hour
  relation of, to kilometers per
        hour, 19
milli-
  meaning of, 5, 9, 62
Milligrams
  relation to grams, 10-11
Milliliters
  relation to
    cubic centimeters, 38, 41
    fluid ounces, 39, 41
    grams of water, 44
    liter, 38, 40
Millimeters
  relation to
    centimeter, 14, 21
    meter, 14, 21
ml, meaning of, 5
mm, meaning of, 5

μ, meaning of, 62

n, meaning of, 62
nano-
  meaning of, 62
nanometer, 63
nanosecond, 63
Ounces, fluid
  confusion of, with weight
        ounces, 39

relation to
    cubic centimeters, 39, 41
    milliliters, 39, 41
Ounces (weight)
  confusion of, with fluid ounces, 39
  number in a pound, 3
  relation to grams, 4, 6, 10

p, meaning of, 62
Parallelogram
  calculation of area of, 73
Photography
  focal length of camera lenses
        in, 17
  use of metric system in, 17
pico-
  meaning of, 62
Pounds
  relation to
    grams, 10
    kilograms, 4, 6, 10
    long ton, 4
    short ton, 3
Powers of 10
  relation to moving of decimal
        point, 64-65,
  use of, in working problems,
        92-98
Prefixes, metric
  complete table of, 61-63
  meaning of centi-, milli-, and
        kilo-, 5, 9, 62
Problems
  conversion, to be solved by
        ratio method, 90-91,
        111-112

easy way to solve, 82–91
in converting Celsius
    thermometer readings
    to Fahrenheit, 54,
    109–110
relating to
    conversion of volumes to
        weights, 48–49, 108–109
    area of triangle, circle, and
        parallelogram, 76, 110
    areas, 31–33, 104–107
    lengths, 22-23, 102–104
    volume of sphere and
        cyclinder, 76, 110–111
    weights, 11–12, 99–102
    use of powers of ten in solving,
        92–98
Pyramid
    calculation of volume of, 75

Quarts
    relation to liters, 37, 41

Rectangle
    calculation of area of, 73

Sphere
    calculation of surface area of,
        74
    calculation of volume of, 75
    liter, diameter of, 37
    liter, surface area of, 37
Square
    calculation of area of, 73
Square centimeters
    calculation of area in, 25, 30
    meaning of, 25, 30

relation of,
    to square inch, 25, 31
    to square meter, 30
surface area of liter
    cube in, 36–37
    sphere in, 37
Square inches
    relation to square centimeters,
        25, 31
Square kilometers
    calculation of area in, 28, 30
    relation of
        hectares to, 28
        meters to, 28, 30
    relation to square miles,
        28, 31
Square meters
    calculation of area in, 27, 30
    meaning of, 30
    relation to
        arcs, 27
        hectares, 27
        square centimeters, 30
        square yards, 26, 31
Square miles
    relation to square kilometers,
        28, 31
Square Yards
    relation to square meters,
        26, 31

T, meaning of, 61
t, meaning of, 5
Table, conversion
    English to metric, 77–81
    metric to metric, 64–68

Temperature
  body, 52
  conversion of Celsius, to
    Fahrenheit, 50–54,
    complete table, 69–71
  how to interpret Celsius,
    without conversion chart,
    52–54
  water boils at, 51
  water freezes at, 51
tera-
  meaning of, 61
Thermometer
  Celsius, 50
  Centigrade, 50
  Fahrenheit, 50
Ton, long
  relation to pounds, 4
Ton, metric
  how to convert to kilograms,
    7
  relation to
    kilograms, 4
    short ton, 4, 6, 10
Ton, short
  relation to
    metric ton, 4, 6, 10
    pounds, 3
Triangle
  calculation of area of, 73

Units
  consistency of, 17–18, 28–29
  use of, in working conversion
    problems, 82–91

Volumes, metric
  answers to problems relating
    to, 107–108
  calculation of, 38–39
  conversion of, to weights,
    using density, 45–46
  measurement of irregular, 40
  problems of converting, to
    weights, 48–49, 108–109
  problems relating to, 41–42
  of sphere and cylinder, 76,
    answers, 110–111
  of sphere, cylinder, etc.,
    72–76
  relation to weights, 43–49

Weights, English
  key conversions of, to metric
    weights, 10
Weights, metric, 3–12
  answers to problems relating
    to, 99–102
  conversion of, to volumes,
    using density, 47–48
  converting from one to
    another, 7–8
  key conversions of, to English
    weights, 10
  problems of converting, to
    volumes, 48–49, 108–109
  problems relating to, 11–12
  relation to volumes, 43–49

Yard
  relation to meter 13, 21